THE METHANE AGE

The GeoJournal Library

Series Editor: WOLF TIETZE

The Methane Age

Edited by

T. H. Lee, H. R. Linden,
D. A. Dreyfus and T. Vasko

KLUWER ACADEMIC PUBLISHERS

DORDRECHT / BOSTON / LONDON

INTERNATIONAL INSTITUTE FOR APPLIED SYSTEMS ANALYSIS

LAXENBURG, AUSTRIA

Library of Congress Cataloging in Publication Data

The Methane age / edited by T.H. Lee ... [et al.].
p. cm. -- (GeoJournal library ; v. 9)
ISBN 9027727457
1. Methane--Congresses. I. Lee, T. H. II. International Institute for Applied Systems Analysis. III. Series.
TP359.M4M48 1988
662'.669--dc19 88-6607
CIP

ISBN 90-277-2745-7

Published by Kluwer Academic Publishers,
P.O. Box 17, 3300 AA Dordrecht, The Netherlands.

Kluwer Academic Publishers incorporates
the publishing programmes of
D. Reidel, Martinus Nijhoff, Dr W. Junk and MTP Press.

Sold and distributed in the U.S.A. and Canada
by Kluwer Academic Publishers,
101 Philip Drive, Norwell, MA 02061, U.S.A.

In all other countries, sold and distributed
by Kluwer Academic Publishers Group,
P.O. Box 322, 3300 AH Dordrecht, The Netherlands.

Printed in The Netherlands

This book is dedicated to the late Ed Schmidt, our friend and collaborator. Not a man of many words, Ed always had a kind word for those around him. He was a modest man, almost spartan in his ways, never sparing himself, particularly when others needed his help. All of us at the Institute will treasure his memory.

THE INTERNATIONAL INSTITUTE FOR APPLIED SYSTEMS ANALYSIS

is a nongovernmental research institution, bringing together scientists from around the world to work on problems of common concern. Situated in Laxenburg, Austria, IIASA was founded in October 1972 by the academies of science and equivalent organizations of twelve countries. Its founders gave IIASA a unique position outside national, disciplinary, and institutional boundaries so that it might take the broadest possible view in pursuing its objectives:

To promote international cooperation in solving problems arising from social, economic, technological, and environmental change

To create a network of institutions in the national member organization countries and elsewhere for joint scientific research

To develop and formalize systems analysis and the sciences contributing to it, and promote the use of analytical techniques needed to evaluate and address complex problems

To inform policy advisors and decision makers about the potential application of the Institute's work to such problems

The Institute now has national member organizations in the following countries:

Austria
The Austrian Academy of Sciences

Bulgaria
The National Committee for Applied Systems Analysis and Management

Canada
The Canadian Committee for IIASA

Czechoslovakia
The Committee for IIASA of the Czechoslovak Socialist Republic

Finland
The Finnish Committee for IIASA

France
The French Association for the Development of Systems Analysis

German Democratic Republic
The Academy of Sciences of the German Democratic Republic

Federal Republic of Germany
Association for the Advancement of IIASA

Hungary
The Hungarian Committee for Applied Systems Analysis

Italy
The National Research Council

Japan
The Japan Committee for IIASA

Netherlands
The Foundation IIASA–Netherlands

Poland
The Polish Academy of Sciences

Sweden
The Swedish Council for Planning and Coordination of Research

Union of Soviet Socialist Republics
The Academy of Sciences of the Union of Soviet Socialist Republics

United States of America
The American Academy of Arts and Sciences

Foreword

The use of methane fuel is expanding rapidly, even in a period of declining world energy consumption, and estimates of world methane reserves outpace current (and even projected) consumption. This methane growth trajectory is coupled with the fact that most current techniques and technologies for finding, drilling, producing, transporting, and converting methane appear to offer ample capacity for growth and improvement.

What explains this situation? First, we must remember that finding and producing methane as a primary energy source, of and by itself, is a relatively recent activity. Previously, most methane was developed as a consequence of finding and producing oil. Second, a host of new activities and methods is emerging to expedite the exploration for and production of gas, such as satellite search (remote sensing), geochemical prospecting, highly steerable drilling packages with hardened electronics, scientific fracturing, and exploration for lithogenic (as well as biogenic) methane, to mention only a few of the exciting developments. The potential combined effect of these new technologies can by no means be fully predicted.

Meanwhile, on the utilization side, the rate of increasing shaft-power conversion efficiency (in combined cycles) continues its steady upward progress: reputable manufacturers now "guarantee" 47% efficiency, 51% has been achieved in prototype operations, and first-class technical university research indicates that conversion efficiency will exceed 60% in a decade or two. The first large-scale methane-to-gasoline plant (based on the Fischer–Tropsch process) is in the shakedown stage in New Zealand. Perhaps more important, several major laboratories are now actively working on direct methane-to-gasoline conversion processes.

The International Institute for Applied Systems Analysis (IIASA) has conducted some background research over the past decade that strongly suggests that methane may fill the gap between the current "world petroleum age" and a potential revival of nuclear power and/or a "clean coal age". The possible emergence of such a world methane system (or several interrelated major national or continental subsystems) rests principally on its availability and acceptability, i.e., methane's environmental advantages and its overall economic viability. These determining factors, of course, will be strongly influenced by the rate and size of the combined advances in the methane-related technologies noted above. The (possible) shift from a petroleum-dominated global fuel system to a methane-dominated system would surely have profound worldwide geopolitical, balance-of-payments, defense and security related, and industrial–structural impacts. We have not yet even identified some primary and many secondary effects.

These preliminary results of IIASA analyses, and the Institute's continuing research interests in energy, technology, and environmental policy issues, culminated in a workshop, held in Sopron, Hungary, involving invited specialists from both East and West, in May 1986. The primary purpose of this high-level, but informal, meeting was to question, criticize, expand, and test IIASA's background hypotheses and other promising research to date. A secondary objective was to identify potentially rewarding issues for further analysis. To what extent we managed to achieve our objectives the reader may judge from this book, based in large part on the selected and edited workshop proceedings.

T.H. Lee
H. R. Linden
D.A. Dreyfus
T. Vasko

Laxenburg, Austria
August 1987

Contents

CHAPTER 1

Energy: Economic and Geostrategic Considerations

Robert A. Hefner, III

1.1. Introduction: The New Economics of Energy

When did recent economic history begin? I believe it began when Neil Armstrong took that "giant leap for mankind" onto the moon. At that moment mankind viewed the Earth from space and gained a shocking new perception – that of our global population living on a small beautiful blue-green ball floating alone in the blackness of space, which Buckminister Fuller first described as "Spaceship Earth". This picture, worth a billion words, reawakened Malthusian fears best expressed by the Club of Rome's popular *Limits to Growth,* out of which grew the political momentum and economic forces that continue to reshape economic systems in contemporary civilization.

These newly awakened perceptions of limited resources and fragile ecosystems led directly to the new liberalism espoused by our youth in the late 1960s, which in turn became a catalyst in the Western democracies for significant increases in social legislation and an incredible growth in governmental regulation throughout the 1960s and 1970s. Unfortunately, the extremely high cost of all that regulation and bureaucracy was not then felt by the consumer/taxpayer/voter owing largely to declining energy prices, which masked the pricetag for those highly inflationary and productivity-limiting mechanisms. This was a time when many people in the West believed that the environment of our planet was being blatantly and unnecessarily fouled by our largely blind focus on economic expansion and growth for growth's sake. Ironically, the political efforts of these concerned citizens exacerbated the problem by adding momentum to the growth of price and regulatory controls and the allocation of large subsidies, the results of which distorted energy production and consumption patterns.

I suspect one of the greatest dangers to a democracy's economic system occurs when government usurps allocation from the market through pervasive energy price and regulatory controls. It is precisely because energy is the most fundamental input to the economic system that government control of the price of energy rapidly distorts the supply–demand balance, which therefore leads to imbalances throughout the entire economic system.

In fact, the cost of energy acts as an accelerator or decelerator on the rate of consumption of all resources. More than 30 years of controls on US and European natural gas prices, and their consequent indirect influence on other fuels, depressed all energy prices during the late 1960s and early 1970s. Thus, consumption of global resources accelerated, without inflation, at the very time that the developed democracies were adding layer upon layer of productivity-reducing and inflation-producing regulations to their systems.

At the same time, this distorted consumption of energy led economists to the wrong (but readily demonstrable) conclusion that energy consumption is inelastic. That growth in GNP requires a proportionate growth in energy inputs to the system was "known" by most all economists and policy makers; therefore, inelasticity in energy consumption became a "known" corollary. Economists, during these falsely comforting years prior to the oil price shocks, built econometric models that accurately forecast increased consumption of energy proportionate to economic growth, so long as there was no major change in the price of energy.

These fundamentally erroneous perceptions of the limits to growth and the inelasticity of energy consumption, coupled with the liberal political majorities of the 1960s and 1970s and their penchant for paternalistic central government controls, resulted in policies and forces that transferred the ability to set energy prices from the consuming democracies to the producers. These anti-market policies gave OPEC the power to effect the largest transfer of wealth ever in history by extracting petrodollars from the consumer constituencies of these liberal policy makers.

In this chapter, I discuss only one sector – natural gas in the USA, the largest energy-consuming economy in the world – and the misconceptions held by most US policy makers that have led to massive energy distortions in the US economic system, distortions resulting in huge financial and productivity losses and the recent inflation–deflation cycle.

1.2. Differences between Oil and Natural Gas Indicating a Large Remaining Natural Gas Resource Base

First, let us focus on how decades of misconceptions regarding the US natural gas resource base have seriously misdirected national energy policies. In order to set the stage for my continuing discussion of natural gas, let me start with a brief summary of my own history.

I am an experienced exploration geologist/geophysicist. I have dedicated my entire career to the search for *natural gas* (unlike my grandfather and father, who spent their lives as independent *oil* men in search of *oil).* I have devoted myself to the research and understanding of (a) the differences between oil and natural gas, (b) development of the natural gas industry (which, I found, did not exist until recently), and (c) the pervasive effects of the energy sector, particularly natural gas, on our economy and national security.

The essence of what I have learned is that natural gas is geologically, technologically, economically, and politically different from oil. Unless we can decouple our natural gas resource projections from those of oil (and oil geology, technology, economics, and politics), we cannot even begin to assess the quantity of remaining natural gas resources. By defining the most significant differences between oil and natural gas, it will be clear how the failure to understand these important differences profoundly distorts most forecasts of energy supply and demand.

1.2.1. Geological difference

Natural gas pervades the earth's sediments and its commercial production is, therefore, not limited by the unique trapping mechanisms necessary for the commercial production of oil. Many of the largest gas-producing areas contain little or no oil. Thus, vast quantities of natural gas can be developed throughout the world where little or no oil is to be found. Because of the pervasive nature of natural gas in sediments, the idea of organizing natural gas statistics by "fields", which is entirely appropriate for oil, has no meaning in natural gas projections. Such miscategorized data have resulted in many erroneously conservative forecasts.

1.2.2. Technological differences

Natural gas is compressible (Boyle's law), oil is not. Oil is a viscous liquid, methane gas is lighter than air. Therefore, gas can be produced commercially from rocks through which oil cannot even move. Moreover, a barrel of oil produced from 20,000 feet below the surface is still a barrel of oil at the surface; whereas a barrel of gas from 20,000 feet can yield as many as 500 barrels of methane at the surface. Thus, as one drills deeper, the quantity of gas recoverable from any given reservoir grows with each increment of pressure, and the effective drainage area of the reservoir available to each bore hole also increases. This fact contradicts the widely held "expert" opinion (used in all econometric resource projections with which I am familiar) that the recovery cost per unit of gas increases exponentially with depth. This lack of understanding prevents deep provinces from inclusion in the resource base, creating a major resource base error that falsely indicates a limited supply.

1.2.3. Political and economic differences

Nearly all of the gas produced and consumed during the past three decades has been subject to stringently enforced low price controls. The result of these low price controls was that virtually all gas was discovered during the search for oil and produced as a by-product of oil. Through decades of price controls at barely a fraction of the cost of oil and other competing fuels, policy makers insured that their prophecies of gas shortages would be self-fulfilling. These controls and regulations distorted US energy supply and demand over such a long period that all statistically based approaches to determining the remaining natural gas resource base are distorted by their economic association with oil. The results are unrealistically and dangerously conservative – more misinformation for the "limits to resources" lobby!

Highly regarded and politically influential natural gas resource studies, such as those by Mehring (1981), Hubbert (1982), and Zareski (1976), do not in any way reflect the true size of remaining natural gas reserves. These and other such *statistical* approaches are useful only as realistic assessments of the undeveloped oil-associated natural gas that remains to be discovered in the future *search for oil*. Therefore, because the "lower 48" is a very mature *oil* (immature natural gas) province, the extremely conservative projections related to *oil* statistics and *oil* understanding are approximately correct indications of oil by-product natural gas. Those statistical studies ignore all relatively undeveloped frontier natural gas provinces in the "lower 48", such as the Anadarko Basin, the Tuscaloosa Trend, the Smackover Trend, the Overthrust Belt, the Arkoma Basin, the Appalachian Province, and other areas. Although these vast natural gas provinces are overlooked in oil-distorted statistics, they *are* included in the most credible geologic estimations made for the USA by the Potential Gas Committee (1983), Parent (1983), Woods and Vidas (1983), and Dolton *et al.* (1981). These more credible projections were made by experts who look forward, i.e., geologists and explorationists, not by those who are expert in looking backward, i.e., statisticians and economists using econometric projections based on the past.

In short, the problem with statistically based estimations, as displayed in *Figure 1.1*, is that they ignore all natural gas that is not associated with oil! A personal anecdote may help to clarify the problem.

During the US Congressional debate leading up to the Natural Gas Policy Act and the Fuel Use Act, which in effect prohibited the use of natural gas, I asked an economist who was affiliated with the American Gas Association why his *optimistic* resource projections for natural gas were, to my mind, so conservative. He replied that he had pushed his calculations to the limits of optimistic projection by using "finding rates" equal to the highest ever achieved in "the industry's" (he meant the *oil* industry's) entire history, "during the mid-1950s". He failed to realize that the entire "industry" in the 1950s was an *oil* industry looking for *oil* in America's *oil* provinces, not a *gas* industry looking for *gas* in America's *gas* provinces.

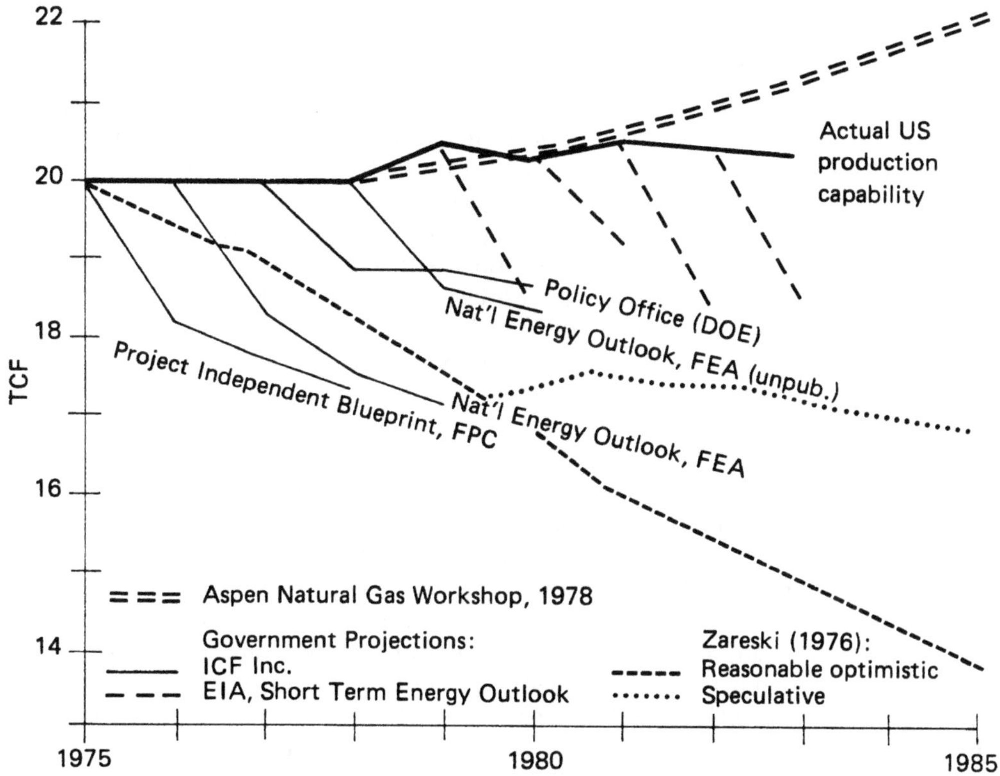

Figure 1.1. Projected versus actual US conventional gas production capability. (Actual production capacity/capability is based on actual production plus excess production capacity as estimated by Goodman, Sachs & Co. oil analysts, April 1984.) (Courtesy GK Companies.)

He was not aware of the fact that all wells drilled during those years were based on risk assessment and financial returns pegged to the *price of oil*, not the *price of gas*. The price of natural gas was not even a factor in the exploratory risk judgment nor in the formation of risk capital during that period.

Still other gas-versus-oil resource distortions are motivated by normal financial considerations. For example, some highly regarded natural gas resource forecasts – many published by the major oil/energy companies and a few natural gas transmission companies – express deep concern over "dangerously limited" remaining reserves. I believe that these distortions, which are generally thought to be realistic or conservative estimates of undeveloped natural gas resources, grow out of "corporate momentum" – that is, institutional memories of the past expenditure of billions of dollars in frontier oil development, oil shale, coal gasification, and the like, all based on the the oil-related model of limited resources, inelastic supplies, and ever-increasing prices. Because the major oil companies must earn the largest possible return from this spent capital for their stockholders, their heavy investments in finding, refining, and marketing oil compel them to view the offshore provinces and the Alaskan oil and gas frontier more optimistically than the "oil mature" lower 48 states. This same logic applies to "gas" companies heavily invested in LNG and coal gasification because, if the existence of an abundant domestic natural gas resource base were proved, their earlier investments would appear to be exceedingly unwise. (I suspect it was El Paso Natural Gas Company's belief in domestic shortages that led its directors to "bet the company" on Algerian LNG ... and lose!) As a consequence, when the major oil companies, the LNG-invested gas companies, and most economists all predicted that the United States was running out of natural gas, how could a policy maker believe otherwise? This general perception of shortages, combined with the significant shift toward the political left throughout the 1960s, thus led to the imposition of federal government controls, which created the economic forces that transferred oil pricing power to OPEC.

Let the record show that I do not condemn econometric statistical approaches to resource projection *per se*. I am simply and emphatically stating that, in the case of North American natural gas reserves, the only relevant statistics are those gathered since 1977, when significant capital formation and risk judgment began to be based on the wellhead price of natural gas, not oil, and focused on the exploration and development of natural gas from natural gas geological provinces, not oil from oil provinces. Although the "statistical indications" since 1977 tend to support strongly the various "geologically oriented" and "optimistic" range of assessments of remaining natural gas, these data have largely been overlooked, owing to the long time lag for inclusion in the public record. In the end, the best evidence of an ample gas supply is the post-1977 natural gas drilling boom, which created today's unrelenting domestic natural gas surplus.

For these reasons, including my own intuition based on a professional lifetime devoted to the search for and the understanding of natural gas, I predict that the remaining US natural gas resource base falls between 1,000 and 1,500 tcf and that North American reserves are at least double: 2,000–3,000 tcf. Thus, natural gas supplies are more than ample to take us at least to the middle of the next century when, I believe, we will move to a nuclear-generated hydrogen economy.

1.3. Transition from an Oil-Based to a Methane-Based Economy

1.3.1. Oil as a mature technology

Precisely because natural gas continues to be linked to the oil industry, which is a mature, fully integrated industry involving a mature energy technology in terms of logistic substitution (Marchetti and Nakicenovic, 1979), the future for natural gas (and, therefore, the necessity for accurate resource projections) continues to be thoroughly misjudged. The facts are that natural gas continues to be globally abundant and, therefore *domestically* abundant in nearly all countries at affordable costs. Natural gas is also free of

(1) The external costs of defense systems necessitated by the geographic concentration of oil (60% of the world's oil reserves are located in less than 1% of the land area).
(2) The terrible pollution costs of coal or high economic cost of "clean" coal.
(3) The political and economic costs of nuclear power.

The picture that emerges by decoupling natural gas from oil is that of a youthful technology that is most likely to fuel all efficient twenty-first-century economies. Such a picture clearly emerges from the logistic substitution work being done at IIASA and the market's continuing progression toward an energy source containing increasing quantities of hydrogen (Lee *et al.*, 1986).

The lower 48 states and offshore areas – and, increasingly, the world at large – are maturing as oil provinces. Oil reserves are declining along with their associated reserves of natural gas. The spate of mergers in the oil industry, reflecting the efforts of major oil companies to wring dry their oil reserves rather than find new fields, reminds me of the desperate actions of a species approaching extinction. These industry efforts clearly indicate the difficulty of maintaining present levels of *oil* production and consumption (Samuelson, 1984). On the other hand, natural gas is emerging as an integrated energy technology from a past during which it was produced merely as a by-product of the oil industry and distributed to consumers through stringently regulated utilities with little regard for commodity costs or fuel value.

1.3.2. The oil analogy

Let us for a moment return to history and use the relevant statistics and lessons of *oil* to understand the present and project the future. Look back to 1925, when most expert forecasters of oil reserves were predicting there could never be sufficient oil to fuel the then correctly perceived growth of a great American industrial economy. At the same time, many renowned geologists professed that they would "drink all the oil" ever found in vast areas of Texas and Oklahoma, now filled with giant oil fields. Look back to 1928 and 1929, when the wellhead price of oil rose to $1.34/barrel and brash men, such as Phillips, Sinclair, Getty, and others, rejected such pessimism because, at $1.34/barrel, the risk of exploration in their minds was justified by the potential for great profits. Look back to the Crash of 1929 and the ensuing Depression that was not felt immediately in the West because, by 1932, the fantastically prolific Oklahoma City and East Texas fields and others were in frantic boom stages of development. Within the next two years, those fields gushed an enormous "glut" of oil into the depressed national and global economies. When the price of oil fell to 10 cents and then to 6 cents per barrel, Oklahoma's Governor "Alfalfa" Bill Murray proclaimed that, "Economic waste is conducive to physical waste," and called up Oklahoma's National Guard to blockade Oklahoma's oil fields and enforce a new concept of conservation through prorationing.

Out of this economic and political chaos, economic forces and the will to survive, not omnibus oil legislation, necessitated and gave birth to a newly integrated and increasingly efficient oil industry. Thus, just as the new vertically integrated "oil industry" emerged as a better and more efficient energy technology to replace abundant coal and wood and to fuel America's recovery from the Depression, an efficient, vertically integrated methane industry is emerging today, six decades later. The natural gas drilling boom of 1978 to 1982 resulted in a supply glut, which depressed demand, causing prices to collapse from $10/unit to $1/mcf. These recent and continuing natural gas industry problems have created the forces, as with oil in the early 1930s, compelling the restructuring and vertical integration of the natural gas industry. From this chaos of the early 1980s, new leadership and direction will emerge to create tomorrow's "Exxons" of the natural gas industry. Natural gas is becoming a better energy technology and will replace oil as the premier energy to fuel America's next great economic pulsation in the twenty-first century. In energy the Kondratieff cycle appears to be frighteningly accurate!

Economic history clearly demonstrates that new technology emerges from periods of severe adversity. Human ingenuity proliferates during the emergence of a new industry and leads to exponential growth in the efficiency of "in hand" technology equaling many orders of magnitude. Recognizable analogies are the automobile industry from the model T to today's best cars, and the aviation industry from the DC3 to the 767.

Today methane technology is roughly equivalent to aviation technology in the 1930s. In its first 30 years the aviation industry accomplished, without revolutionary scientific breakthroughs, a productivity gain of roughly 100 times, if one compares the DC3 to the 747 (Nakicenovic, 1986). Thus, as we have experienced in the global development and proliferation of the aviation industry, the first few decades of a new and emerging industry are likely to increase the productivity of "in hand" technology sufficiently to bring forth enormous new markets and unforeseen horizons. The natural gas industry is set to make equivalent productivity gains and to feel the effects of evolutionary and revolutionary improvements in methane technology. Technological improvements and productivity gains will occur in exploration, drilling, production, pipelining, distribution, ocean transport, and conversion to electricity (high-efficiency combined cycle gas turbines). Additionally, direct conversion of methane to competitively priced, clean liquid fuels holds enormous implications for the future of the world's transportation sector. Simply stated, *as methane technology improves, the amount of methane available for reserves and resources will increase exponentially* (Osborne, 1984).

1.3.3. The natural gas glut

Despite all the promising benefits of an emerging methane industry and expanded natural gas usage, exploration and development in the USA have nearly stopped. The pessimistic resource estimates of the past and rigid price regulation combined to create a self-fulfilling prophecy of natural gas shortages, resulting in plant and school closings and other economic disruptions during the winters of 1977 and 1978. Such shortages and the resulting distorted natural gas "corrective" legislation led directly to the gas drilling boom of 1970–1982. These misperceptions of shortages, as exemplified by "Club of Rome thinking" and fears about finite resources, the oil shocks of the 1970s, and predictions of $100/bbl oil drove gas prices beyond economic reality. Runaway hyperinflation of 30% to 50% per year seized the gas drilling industry, and an equally damaging decline in efficiency occurred because of overtaxed equipment and inexperienced personnel. Demand declined rapidly due to rising prices, "new found" elasticity, a deep recession, and a shift away from natural gas to fuel oil by heavy industry in the USA. Meanwhile, the natural gas resource base was being vigorously developed, leading to enormous (and surprising to most experts) increases in the producing capacity of domestic natural gas and to what now is called a continuing natural gas "glut". This sudden surplus of natural gas, combined with diminishing demand, led to falling wellhead prices and *de facto* deregulation. The boom was over. The bust is in its fourth year, and the US natural gas industry is suffering severe economic pain, resulting directly from 30 years of distorting regulation of gas as an oil industry by-product.

The gas glut continues, while the USA continues to suffer unnecessarily from the economic and national security problems of increasing oil imports, coal's acid rain destruction, and large contributions to budget deficits and higher electricity costs by the nuclear sector. The deteriorating demand for natural gas relates principally to the lack of a clear national policy to release the shackles of regulation and promote competition in the transportation and end-use sectors. Natural gas drilling in the USA remains at dangerously low levels, which once again will fulfill prophecies of shortages in the midst of an abundant domestic resource.

Natural gas demand in the USA is limited by:

(1) Massive irrational transportation and distribution regulation.
(2) Unnecessarily high end-use prices caused by (1) above.
(3) Loss of large quantities of demand in US heavy industries (autos, steel, chemical) as they grew less competitive with their foreign competitors.
(4) Unfortunate and untimely federal and state laws and regulations prohibiting, limiting, or discouraging the use of natural gas.
(5) Fears of future shortages and their economic disruption, of more legislation, of incremental pricing, and the totally unwarranted fear of a limited North American resource base.

1.4. The Nuclear Non-option

I believe that the most fundamental aspect of an efficient, competitive, and secure economy is access to an abundant, moderately priced, and dependable source of energy. Because production of all goods and provision of all services require the expenditure of energy, energy is the essential input for the expansion of any economic system and improvements in the quality of life.

Globally, we seem trapped between the necessity to insure an abundant and adequate supply of energy to revitalize the developed nations and retire the debts of the less developed countries *versus* the potential devastation of the nuclear option. Fortunately, our spaceship Earth offers an alternative – natural gas – a socially and environmentally benign, comparatively inexpensive, abundant, and ubiquitous supply of energy, awaiting development by both the developed and developing economies.

To those who argue that "For many, perhaps most, countries there is at present no alternative to nuclear energy" (Akhund, 1983), I would respond that the opposite is true: for many, and perhaps most, countries, methane and methane-generated electricity can become a significant energy source meeting all demand for new increments of energy as needed at substantially lower costs than nuclear power.

With all the current economic and debt-related problems of the non-oil-producing less developed countries (LDCs), it would be both economically inefficient and geostrategically imprudent for the Western nuclear nations to foist onto these weaker, already debt-ridden economies a capital-intensive and technologically inefficient system for the production of energy. It is now obvious in the USA as well as most of the OECD countries that energy demand is in fact very "elastic", that natural gas is abundant, and that nuclear energy never was a "necessity". The United States, perhaps out of guilt for dropping the atomic bomb (as expressed by our "Atoms for Peace" program) as well as our policy makers' misperception of shortages and finite resources typical of Club of Rome thinking, invested well over $1,000 billion and hundreds of thousands, if not millions, of man-years in a sadly inefficient, capital-intensive nuclear system. Hindsight proves that this investment was neither necessary nor efficient – a sad waste of capital and human talent and even more tragic should the LDCs decide to repeat the error. The non-oil-producing (but often gas-prone) LDCs must not be trapped by the same outdated thinking and must beware of the large institutional momentum for the sale, export, and financing of nuclear plants by those who invested enormously in nuclear technology, plant, and equipment. Nuclear proliferation is undoubtedly the Achilles heel in revitalizing the international order and will continue to be so, as certainly as night follows day, unless we can break the perception held by many of the world's enlightened leaders that there is no alternative to the continued development of nuclear power.

Natural gas is undoubtedly the emerging alternative, and I predict that its decoupling from oil will lead to globally beneficial growth well into the twenty-first century.

References

Akhund, I. (1983), *On Revitalizing the International Order* (Aspen Institute for Humanistic Studies, Aspen, CO).

Dolton, G.L. *et al.* (1981), *Estimates of Undiscovered Recoverable Conventional Resources of Oil and Gas in the United States*, Circular 860 (US Geological Survey, Washington, DC).

Hubbert, M.K. (1982), Techniques of prediction as applied to the production of oil and gas, in S.I. Gass (ed.), *Oil and Gas Supply Modeling*, Special Publication 631, May (National Bureau of Standards, Washington, DC).

Lee, T.H., Schmidt, E., and Anderer, J. (1986), *Natural Resources and Energy Systems: A Strategic Perspective*, Working Paper WP-86-030 (International Institute for Applied Systems Analysis, Laxenburg, Austria).

Marchetti, C. and Nakicenovic, N. (1979), *The Dynamics of Energy Systems and the Logistic Substitution Model*, Research Report RR-79-013 (International Institute for Applied Systems Analysis, Laxenburg, Austria).

Mehring, R., with E.R. Van Driest II (1981), *The Discovery of Significant Oil and Gas Fields in the United States*, R-26541/1DOE (Rand Corporation, Santa Monica, CA).

Nakicenovic, N. (1986), Transportation and Energy Systems in the United States (Draft), National Academy of Engineers Workshop, August 13, Woods Hole, MA.

Osborne, D. (1984), America's plentiful energy resource, *Atlantic Monthly*, March, 86–102.

Parent, J.D. (1983), *A Survey of United States and Total World Production, Proved Reserves, and Remaining Recoverable Resources of Fossil Fuels and Uranium* (Institute of Gas Technology, Chicago, IL).

Potential Gas Committee, Potential Gas Agency (1983), *Potential Supply of Natural Gas in the United States* (as of December 31, 1982), June (Colorado School of Mines, Golden, CO).

Samuelson, R.J. (1984), Economic perspective: Merger debate reeks of fuzzy thinking, *Washington Post*, March 24.

Woods, T.J. and Vidas, E. H. (1983), *The GRI Hydrocarbon Model: The Structure, Results, and Sensitivities of the Model in its Initial Form*, May (Gas Research Institute, Chicago, IL).

Zareski, G.K. (1976), Chief of Planning and Development Division, Bureau of Natural Gas, Federal Power Commission, statement from *Long Term Prospects for U.S. Gas Supply*, Hearings before the Subcommittee on Energy and Power of the Committee on Interstate and Foreign Commerce, US House of Representatives, January 27.

CHAPTER 2

The Dynamic Evolution of Methane Technologies

A. Grübler and N. Nakicenovic

2.1. Introduction

The use of natural gas has been increasing in many parts of the world, especially in most industrialized countries. As a result, natural gas has emerged as one of the three most important sources of energy, ranking third in the world after oil and coal, while in the USA it is second only to oil. Despite enormous increases in natural gas consumption and related improvements in production, transport, conversion, distribution, and end-use technologies, natural gas is still considered the "stepchild" of the oil industry – a by-product of oil production.

This is the more surprising considering the promising prospects for the widespread use of natural gas in the future. Natural gas is cleaner than any other fossil energy source, and unlike other fossil energy forms, produces limited particulate and sulfur emissions that can be even further reduced by relatively simple measures. In the past, most natural gas discoveries were incidental to oil prospecting. Today, there is increasing evidence that methane may be more abundant and distributed more evenly throughout the world than other fossil energy sources. In fact, estimates of natural gas resources have increased substantially during the last decade. However, despite decisive environmental advantages and a potentially abundant supply, the use of natural gas stagnated during recent years. The gas bubble still persists because acquiring new markets turned out to be more difficult than had been anticipated by the promoters of natural gas during the phase of rapid growth that lasted until a few years ago.

Our contention is that most of the difficulties encountered in attempts to increase the use of natural gas could be resolved if specific technologies were to be developed that are tailored to natural gas and are not mere derivatives of oil technologies. In other words, a prerequisite for the widespread use of natural gas in the future is the increasing decoupling of methane technologies from oil technologies. As a consequence, the oil and gas industry would eventually split into two separate entities. Before embarking on the justification of our contention, we first describe the evolution of the energy system and the dynamics of natural gas. With this historical perspective, we then outline the future developments likely to lead to the separation of the oil and gas industry, and of oil and methane technologies.

2.2. Primary Energy Consumption

At the beginning of the nineteenth century, the primary energy inputs were fuel wood, agricultural wastes, and mechanical wind and water power, in addition to animal and human muscle power. Poor as this may be, by present standards, it represents a sophisticated system compared to earlier practices. A considerable infrastructure of canals and roads was already in place for timber transport; mining and manufacturing were usually associated with elaborate systems of dams and water-wheels; thus, motive and shaft power came from draft animals and hydraulic systems, and heat from biomass. In primary energy terms, fuel wood represented most of the energy inputs.

Figure 2.1 (Nakicenovic, 1984) shows primary energy consumption in the world since 1860. The data are plotted on a logarithmic scale and show exponential growth phases in consumption of the most important sources of primary energy during the last 130 years in piecewise linear secular trends. Consumption of fuel wood, once the most important source of energy, has declined since the beginning of the century, although its use is still widespread, especially in the developing parts of the world. With the expansion of railroads and the steel industry and the application of steam in general, coal use increased exponentially until the 1910s and has oscillated ever since with an overall lower average growth rate. Both oil and natural gas were introduced during the 1870s, and their consumption has increased, with even more rapid exponential growth rates ever since. In fact, the oil and natural gas curves have the same shape and almost identical growth rates; they are just shifted in time by about 10 to 15 years. Oil and natural gas use grew in parallel with the petrochemical industry, the electricity and electrical industry, internal combustion and electric prime movers. Nuclear energy is still in its early phase of development; therefore the steep growth rates prevailing over the last decade may not be indicative of its future role. During recent years, the growth of nuclear energy has declined worldwide to more moderate rates.

Thus, during this 130-year period, energy consumption did not draw equally from all sources, nor did the use of all energy sources increase equally. Yet, global primary energy consumption (including fuel wood) grew exponentially at an average rate of 2.3% per year. It is evident that the

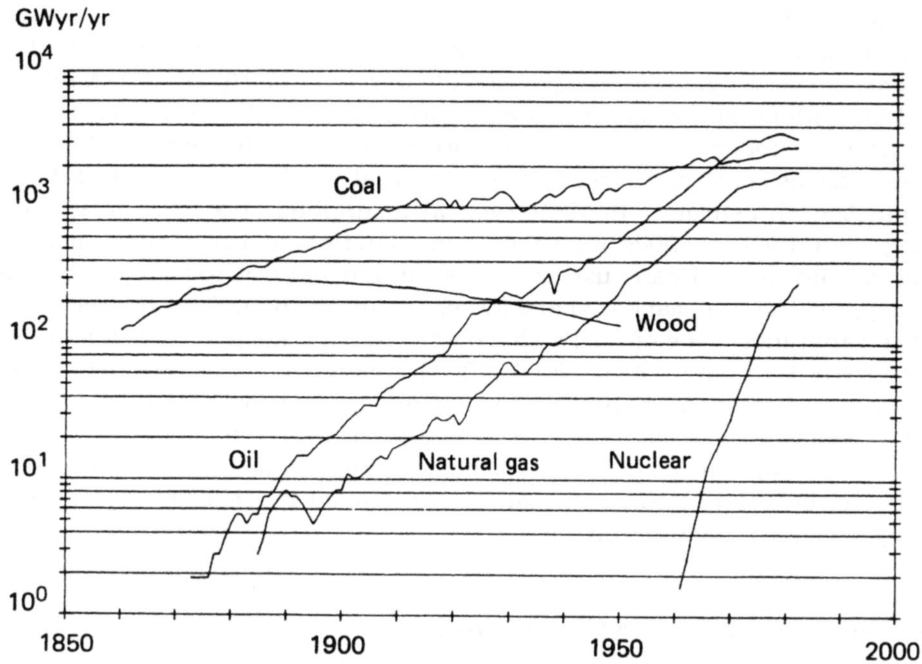

Figure 2.1. World primary energy consumption.

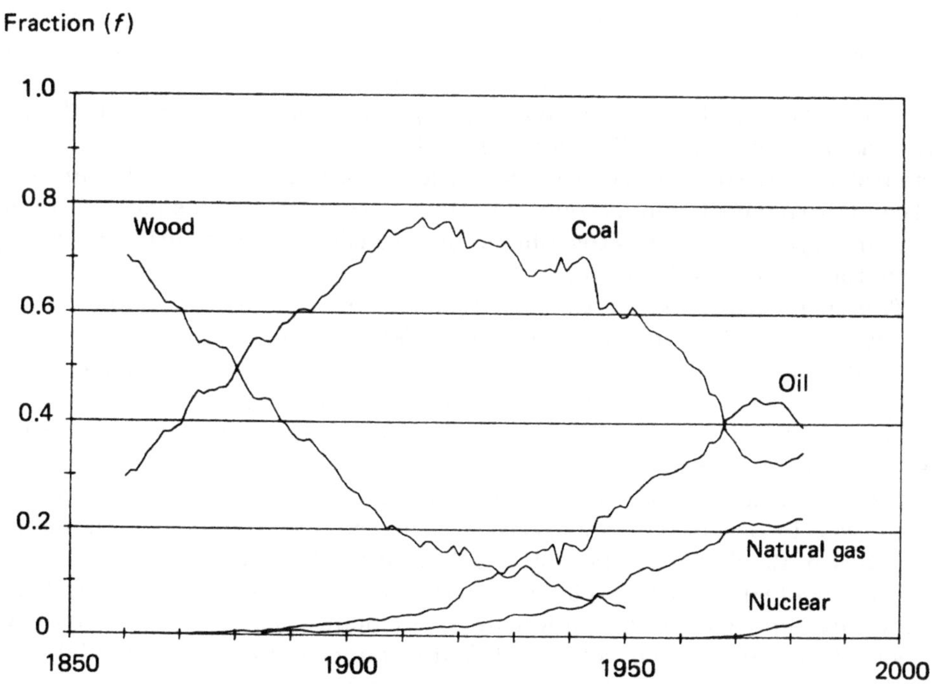

Figure 2.2. World fractional shares of major primary energy sources.

older forms of energy have been replaced by new ones. Thus, the decline of the older energy sources was compensated by the more rapid growth of the new ones. These dynamic changes are clearly seen in *Figure 2.2* (Nakicenovic, 1984), which shows the fractional market shares of the five most important primary energy sources taken from *Figure 2.1*. In terms of fractional market shares, coal had already replaced fuel wood during the last half of the nineteenth century. In 1860, fuel wood supplied about 70% of consumed energy, but by the 1900s its share had dwindled to little more than 20%. Owing to the insignificant use of crude oil and natural gas during the last century, most of the market share losses incurred by fuel wood were caused by the rapid increases of coal's share of primary energy – from 30% in 1860 to almost 80% by the 1900s. By 1910, the rapid increase in coal use had ceased, and during the 1920s, a phase of decline set in. This decline in the relative share of coal use resembles the market losses of fuel wood 50 years earlier. The replication of this pattern is almost symmetrical because, after the 1920s, both fuel wood and coal were replaced by still newer sources of energy – crude oil and natural gas.

2.3. Natural Gas in the Global Context

The evolution of primary energy use, viewed as a technological substitution process, is shown in *Figure 2.3* (Nakicenovic, 1984) on a logarithmic plot of the fractional market shares of the five primary energy sources (from *Figure 2.2*). The fractional shares (f) are not plotted directly, but rather as the quantity $f/(1-f)$ that transforms the logistic curve into a straight line (i.e., as the linear transformation of the logistic function). The quantity is the ratio of the market share taken by a given energy source over the sum of the market shares of all other competing energy sources. This form of presentation reveals the logistic substitution path as an almost linear secular trend with small annual perturbations. Thus, the presence of linear trends in *Figure 2.3* indicates where the fractional substitution of energy sources follows a logistic curve.

The model estimates of the substitution process are extended beyond the historical period up the the year 2050 [1]. For such an explorative "look" into the future, additional assumptions are required because potential new competitors, such as nuclear and solar energy, have not captured sufficient market shares in the past to allow estimation of their penetration rates. We have assumed a more modest nuclear penetration rate that resembles the historical growth rates of the introduction of coal, crude oil, and natural gas into the energy system. The nuclear scenario, therefore, prescribes a 1% share in 1965, and a 3% share 25 years later in 1990 (in 1982, the nuclear share in global primary energy consumption was 3.5%). For the next energy source, which we symbolically call "solfus" to indicate the potential use of both solar and fusion energy, we have postulated an equivalent scenario with a 1% share in the year 2025, rising to 3% in 2050.

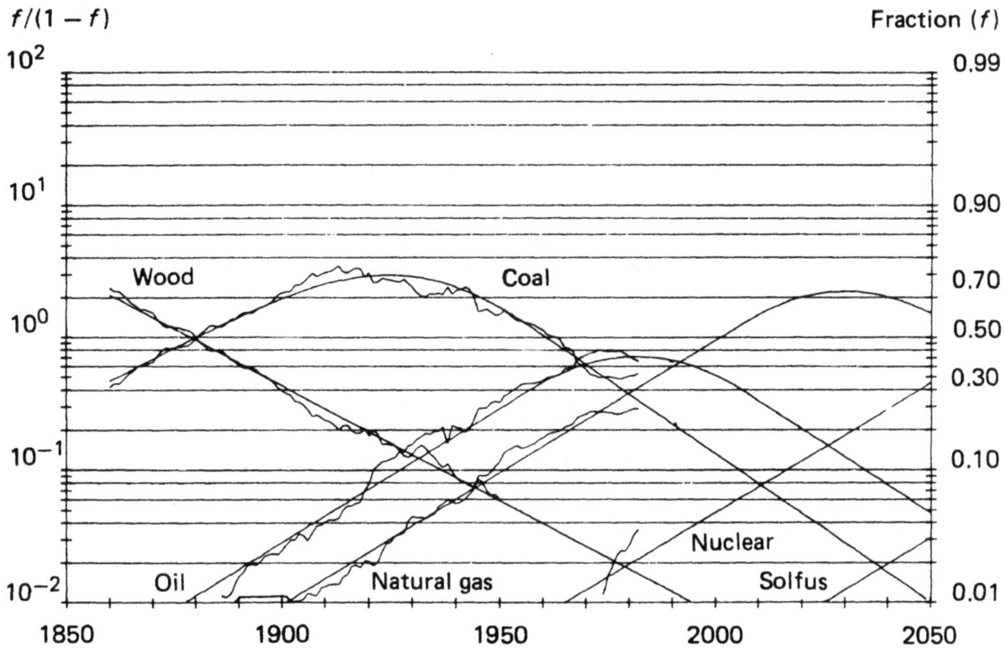

Figure 2.3. World primary energy substitution (with projections).

These two assumptions, together with the dynamics of energy substitution prescribed by past events, describe the resulting evolution of the global energy system throughout the first half of the next century.

The prominent feature in this projection of primary energy substitution dynamics into the future is the emergence of natural gas as the dominant energy source during the next decades. According to *Figure 2.3*, more than half of all the primary energy consumed globally will be natural gas after the end of this century. This result illustrates that not only would the natural gas bubble be absorbed in a few years, but that methane technologies would develop in the future, creating new growth sectors. Although this result is unexpected in terms of the numerous energy debates of the last 10 to 15 years, it is perhaps reassuring that we may not, after all, have to rely on nuclear or alternative energy sources for another 50 years or so. Instead, the possible future that emerges would require less radical changes, but we still face a challenging task to develop new technologies and to improve the performance of those already employed, such as deep drilling, pipelines, and methane conversion into other energy carriers (e.g., electricity and methanol). Despite the current difficulties involved in expanding the use of natural gas in a time of worldwide economic slowdown and low energy (i.e., crude oil) prices, our scenario paints a different picture, albeit only in the long run.

To gain a better understanding of how such changes may come about, we will first investigate the history of natural gas in greater depth. Thereafter, we return to the broader picture of primary energy and the creation of the new growth sectors that could emerge from expanding methane technologies and their use.

Figure 2.4 shows the history of natural gas production for the world. The shares of the four most important producing regions are plotted since 1900. North America (the USA and Canada) was the dominant producer of natural gas until 1983, being superseded by the Soviet Union during the subsequent three years. In fact, the USA has produced most of the natural gas ever extracted globally, and still continued to produce more than half until the mid-1970s. The history of natural gas is, therefore, closely linked to the USA.

Figure 2.4 also shows the model estimates of the actual market shares of the four major producing regions, but they are included for illustrative purposes only and are not intended to indicate likely future development. Rather, they indicate that the actual market shares of the four producing regions fluctuated widely away from the model estimates, especially from World War II through the 1960s. However, the historical trend away from North American dominance and toward a more widely distributed natural gas production throughout the world should continue, and this is one of the

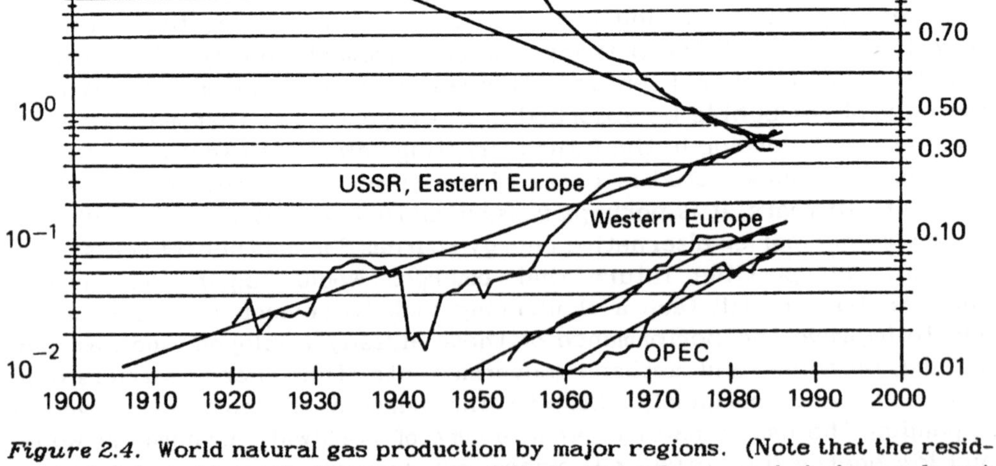

Figure 2.4. World natural gas production by major regions. (Note that the residual to total world production was not plotted. Therefore, market shares do not add to 100%.)

crucial issues associated with the future use of natural gas to which we will return. However, we will next analyze the evolution of the natural gas industry in the USA, since it has been the dominant producer and still is the largest consumer of natural gas, and because US data are readily available.

2.4. Natural Gas in the USA

The USA has a longer recorded history of primary energy use than anywhere else in the world. *Figure 2.5* gives its annual consumption of all fossil energy sources, fuel wood, direct uses of mechanical water power, and hydroelectric power starting in 1800, while *Figure 2.6* (Nakicenovic, 1984 and 1986) shows the substitution among these energy sources. Mechanical power (mostly water and some windmills) and hydropower cannot be seen in the figure due to their low contribution to the total energy supply. They

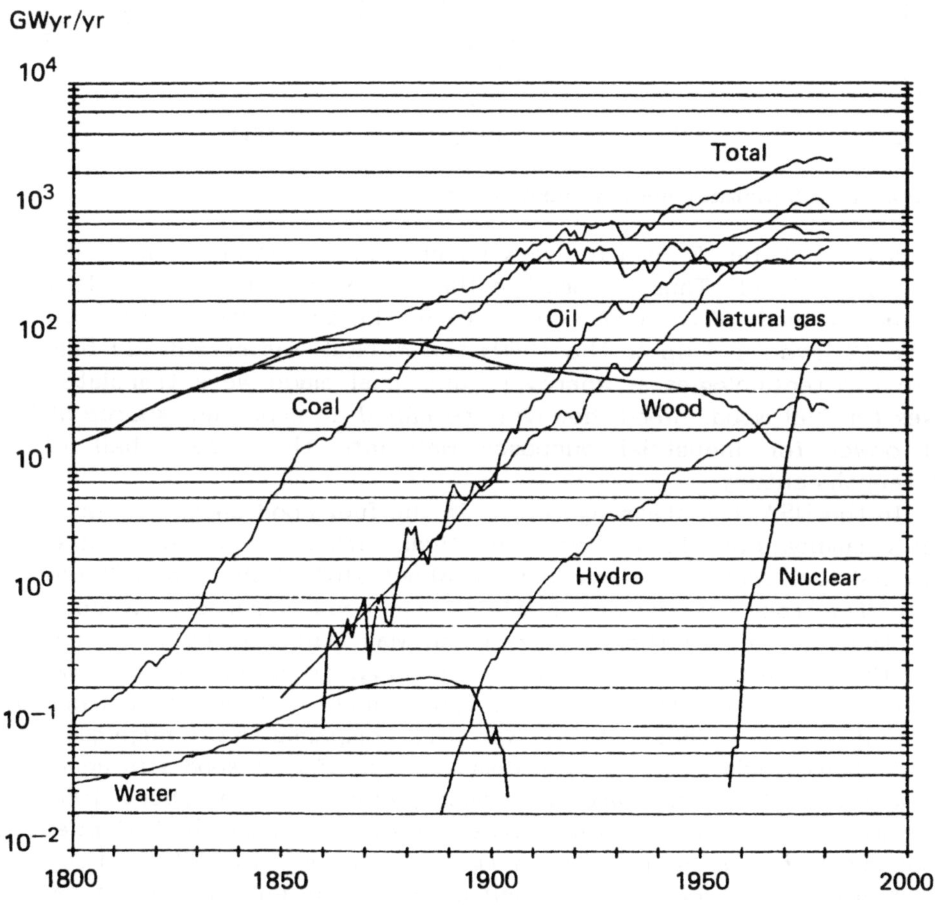

Figure 2.5. US primary energy consumption.

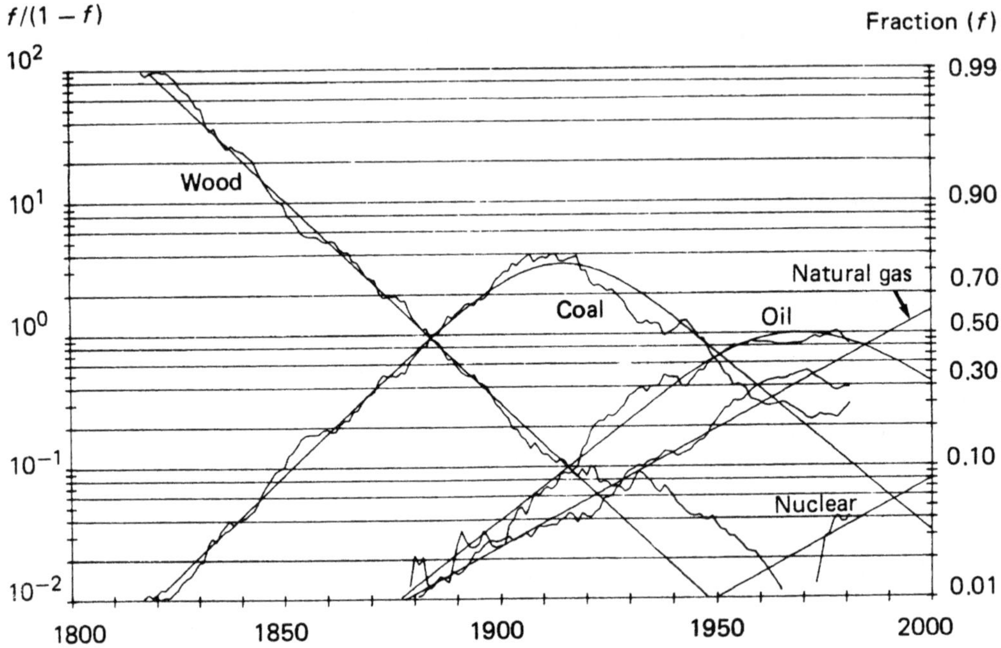

Figure 2.6. US primary energy substitution.

barely exceed the 1% level for very short periods and otherwise fall under that critical level. Thus, before the 1820s, fuel wood provided virtually all the energy needs of the USA. Coal entered the competition process in 1817 at the 1% level, and up to the 1880s the market was essentially based on these two technologies – whatever gains coal made were translated into losses for fuel wood. Wood, however, remained an important source of heat and power for industrial purposes well into the second half of the nineteenth century.

In the USA, the steam age began in the fuel wood-based economy. The first steamboats and locomotives were fired with wood, which remained the principal fuel used by railroads until about 1870 (Schurr and Netschert, 1960). The iron industry was another large wood consumer. Around 1850, more than half of all the iron produced was still smelted with charcoal. Nevertheless, during this early period of industrialization, the USA was still basically a rural society, so that the total amount of fuel wood consumed in manufacturing and transportation was small compared to the huge quantities used in households. In 1880, the domestic use of fuel wood still accounted for more than 96% of fuel wood consumed (Schurr and Netschert, 1960). At the same time, however, coal was already supplying almost half of all energy needs, most of it being used by emerging industries. In 1880, coal supplied almost 90% of the fuel used for smelting iron. The end of the last century therefore marks the beginning of the industrial development period in the USA.

The first use of crude oil and natural gas in the USA dates back to the beginning of the nineteenth century, and during the 1880s it reached the 1% market share. From then on, the use of crude oil expanded somewhat faster as time progressed, and in 1950 crude oil consumption surpassed that of coal. The use of natural gas surpassed that of coal nine years later. It should be noted that, as late as the 1920s, the use of crude oil was not much larger than the consumption of fuel wood.

It is remarkable that the structure of energy consumption changed most during the period of oil dominance. The 1950s – when oil became the dominant source of energy – represent the beginning of more intense competition between various energy sources both in the USA and in the world. For over 150 years whichever energy source dominated the contemporary energy supply also contributed more than half to all primary energy consumption – from 1800 to 1880 this was fuel wood, and from 1880 to 1950 it was coal. During the 1970s, crude oil was close to achieving a 50% share, but before actually surpassing this mark its dominance began to decline. Thus, during the last three decades, three important sources of energy have shared the market, without a single source having overall dominance, which differs from the pattern observed during earlier periods.

Figure 2.6 indicates that, after the 1980s, natural gas would become the dominant energy source, although crude oil would still maintain a roughly 30% market share by the end of the century. Similarly at the global level, future potential competitors of natural gas, such as nuclear or solar energy, have not yet captured sufficient market shares to allow an estimation of their future penetration rates. The starting point for market penetration of nuclear energy can be dated back to the 1960s, when nuclear power acquired slightly less than a 1% share of primary energy. Making allowance for further cancellations of planned power plants and possible decommissioning of those in operation and construction, we have assumed that nuclear energy could at most double its current 4% market share to about 8% by the year 2000. This leaves natural gas with the lion's share of primary energy, advancing its position to the dominant energy source after this century closes.

2.5. Dynamics of Oil and Natural Gas

The earliest historical records of natural gas drilling and use are reported in ancient China, where natural gas was discovered incidentally during the drilling of brine wells (cf. Brantly, 1971; Gaz de France, 1970 and 1971; and Peebles, 1980). These ancient wells were completed with percussion drills and bamboo casings. Some of the oldest wells were reported by Confucius in 600 B.C. to have reached 500 meters and to have produced natural gas that was transported in bamboo pipes for use in evaporating brine to recover salt. By the nineteenth century, this drilling technology improved in performance by almost an order of magnitude. Visitors to China have described drilling depths of up to 4,000 meters, which is comparable to the depths of many commercial wells today. In fact, some of the first natural gas discoveries in the West were also the result of drilling brine wells. Just as in China, the first significant use of natural gas was to dry salt, one of the largest energy consumers among the modest production processes in the pre-industrial age.

2.5.1. Drilling performance and average depth

Modern drilling technology for oil and gas developed originally in the methods used for brine and water wells. The first producing oil well was completed in 1745, in the French Pechelbronn oil field. Drilling technology gradually improved, and by the 1850s, depths of about 600 meters were achieved in France with dry rotary rigs. In the USA, similar depths were reached during the same period with cable tool rigs.

The significant advances in drilling technology since the 1850s were primarily a result of intensive oil drilling, especially in the USA. Probably the most important single innovation in drilling methods was the hydraulic rotary rig, which was introduced around the turn of the century. The dramatic improvement in the performance of drilling technologies during the last 100 years is illustrated in *Figure 2.7*, which shows the drilling depth records for the three most important technologies: dry rotary and

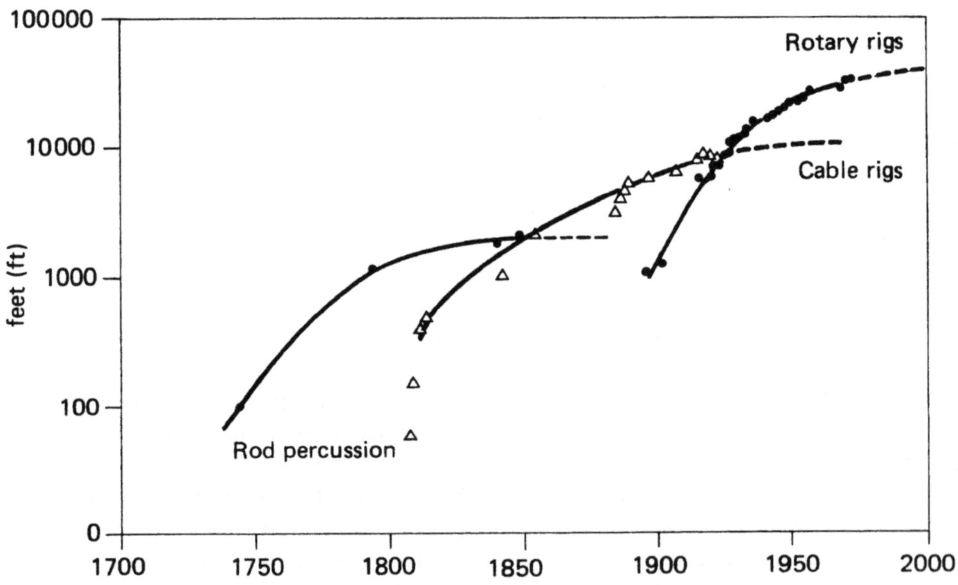

Figure 2.7. World drilling depth records

rod percussion, cable tool, and hydraulic rotary rigs. The data for this figure are taken from Brantly (1971) and *Oil and Gas Journal* (1977). Two small-diameter core test drillings in Germany were excluded from the data, one reaching 2,000 meters in 1893 and the other 2,240 meters in 1909, because they were not exploratory wells for commercial production of hydrocarbons. It can be seen from *Figure 2.7* that both of the new drilling technologies were inferior to the older competitor in terms of record depth at the time of their introduction. In time, however, the new technology overtook the older one to establish and improve the depth records.

In terms of petroleum geology, this trend favors natural gas since the probability of finding methane increases with depth, and that of discovering oil decreases. Between depths of 1,000 meters to a few thousand meters, the hydrocarbon deposits are mostly in the form of crude oil. Thereafter, the likelihood of oil deposits decreases, and below 4,000 meters virtually all hydrocarbon deposits are methane or methane with carbon dioxide (see Donat, 1984). Geological evidence and the chemical characteristics of hydrocarbon compounds indicate a very low probability for the occurrence of more complex molecules (crude oil) below these depths and an increasing probability of finding methane at greater depths due to high pressure and temperature (see Gold, 1985).

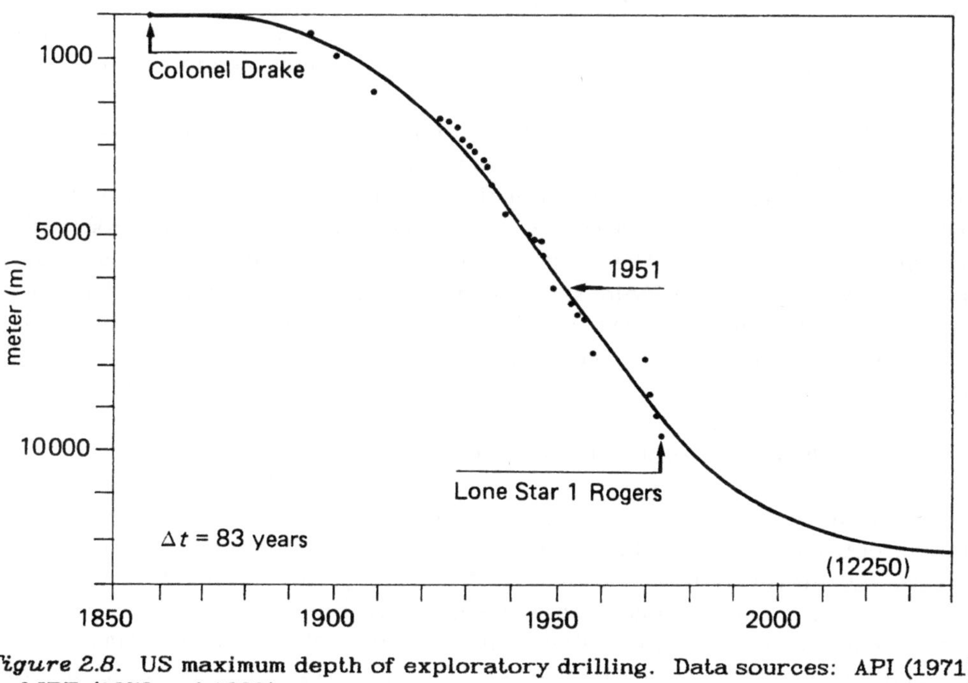

Figure 2.8. US maximum depth of exploratory drilling. Data sources: API (1971) and IPE (1978 and 1983).

Technological progress in drilling, expressed in terms of maximum depth, indicates that new record wells, if successful, can be expected to find methane, not crude oil. In fact, virtually all recent record wells or deep wells were drilled for natural gas exploration. Thus, it is equally interesting to look at the evolution of record depths drilled during the last 100 years for all technologies from *Figure 2.7* taken together. *Figure 2.8* shows the exploratory drilling depth records in the USA, culminating with the Lone Star 1 Rogers, completed during the 1970s, as the deepest exploratory well drilled. This record will most likely be exceeded in the future. *Figure 2.8* shows that the logistic trend in deep drilling would reach an asymptotic depth record exceeding 12,000 meters.

Thus, while the decoupling of oil and natural gas technologies is not reflected in drilling technologies, since these are still basically identical whether the well is drilled for oil or gas, most of the record wells reaching depths below a few thousand meters were primarily natural gas wells in the Anadarko Basin. This tendency toward deeper exploratory wells should favor additional gas discoveries since the probability of finding oil below 10 km is virtually zero. In fact, most of the recent technical improvements in rigs and drillbits are designed for deep wells, both scientific research wells (such as the Soviet effort on the Kola Peninsula) and natural gas prospecting efforts. At the same time, there is some indication that, at greater depths, potentially large unconventional and abiogenic methane deposits may be found (see Gold, 1985). The long-term evolution of drilling technologies to deeper horizons may indeed lead to the overwhelming dominance of natural gas in successful wells, especially as the more readily available oil deposits become exhausted.

In this context, it is interesting to note that these saturation trends in drilling depth records cannot be observed for offshore drillings. *Figure 2.9* shows the worldwide offshore depth records for commercial exploratory drilling (API, 1986; IPE, 1983; and *Ocean Oil Weekly Report*, 1986). The progress achieved since the first so-called oil shock of 1973 is especially noteworthy. *Figure 2.9* indicates that a considerable potential exists for a further increase in the share of offshore gas production, which currently accounts for about 20% of global gas production. It has, in fact, been estimated that up to 40% of the undiscovered oil and gas reserves may be found offshore (Klemme, 1977).

The evolution of exploratory drilling indicates that a long-term trend toward a higher natural gas-to-oil well ratio can be observed for the USA. Although the conventional exploratory drilling technology reached rather shallow depths during the last century (compared with eight or more kilometers today), nevertheless the effect of gradually decoupling methane from oil technologies can be observed. The USA has been chosen for study primarily because of the availability of almost complete historical time series for drilling activities and because of US technological leadership in this field since the beginning of the industry.

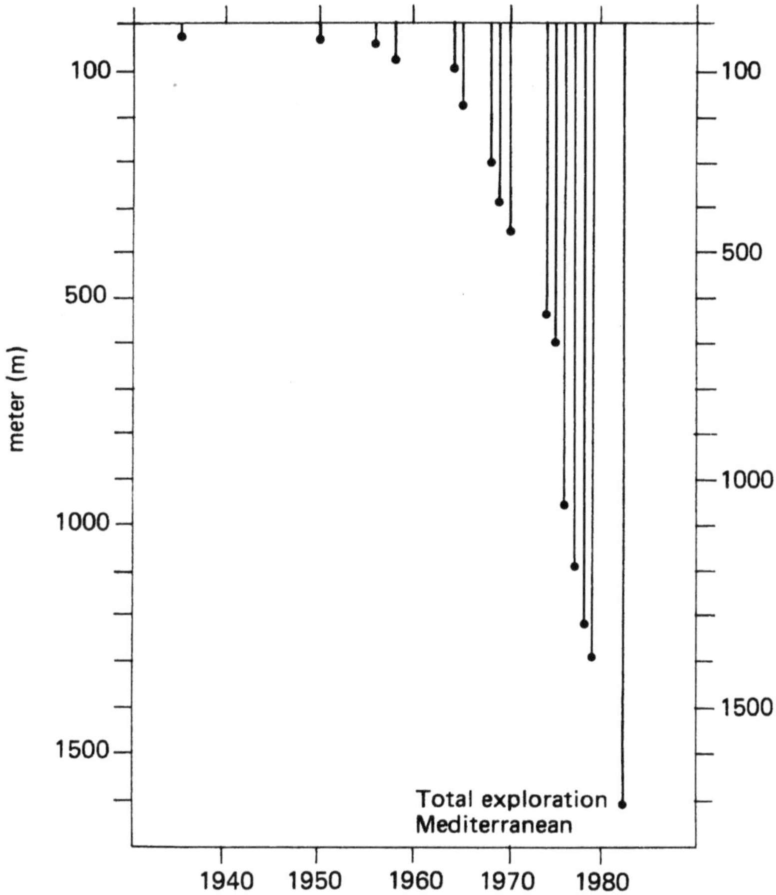

Figure 2.9. **World water depth records in exploratory drilling. Data sources: API (1986) and IPE (1983).**

Figure 2.10 shows the total number of exploratory wells drilled in the USA since 1900, subdivided into oil, gas, and dry wells (API, 1971 and 1986). Alternative drilling statistics have been compiled by the USGS and published in the yearly issues of *Mineral Resources of the USA*. These statistics indicate a much greater number of gas wells drilled, especially in the early period. For instance, the cumulative number of gas wells drilled up to 1908 is reported by the USGS as being 21,300, whereas the API statistics give a cumulative total of 3,185 to the same year. Clarification of such differences would require further historical analysis. Service wells are excluded from the drilling statistics used in the examples. Equally, when dealing with these statistics, it is important to note that, during the early period (i.e., before the 1920s), many natural gas finds, discovered in the course of oil exploration, may have been classified as dry holes, due to the lack of a commercial value of a great part of the deposits discovered.

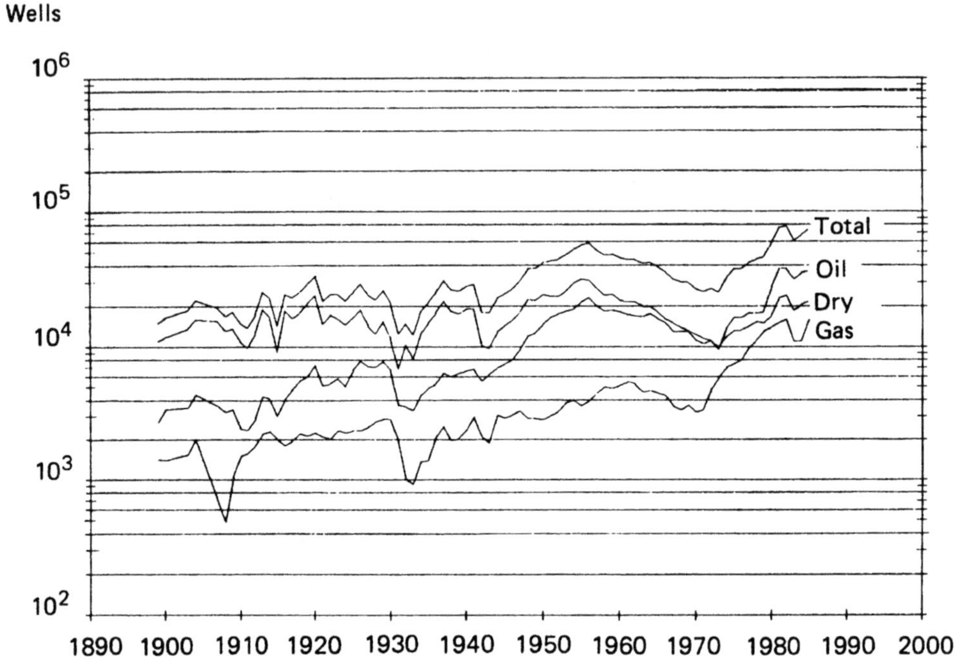

Figure 2.10. USA: number of wells drilled.

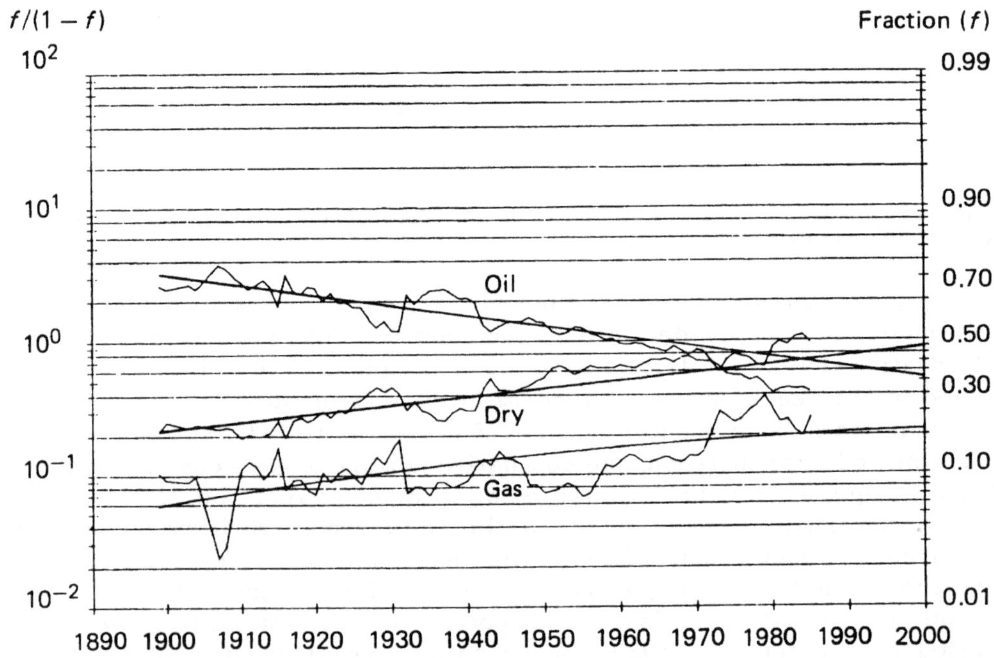

Figure 2.11. US share of oil, gas, and dry wells in total drillings.

Despite these possible shortcoming of the official API drilling statistics, however, *Figure 2.10* indicates a large increase in the total number of wells drilled since 1900, and also large fluctuations, with particularly strong dips in drilling activity during the 1930s and 1960s. *Figure 2.11* shows that these fluctuations disappear, and strong secular trends emerge, when the total drilling effort is viewed as a "market niche" for oil, gas, and dry wells. Defined in this way, the shares of gas and dry wells are increasing and the share of oil wells is declining. Thus, the decoupling of oil and gas technology is manifested implicitly in drilling statistics. Although oil wells still represent the majority of wells drilled, the number of gas wells is on the increase. Unfortunately, this development has been paralleled by an increase of dry wells; but since the late 1960s, there are signs of a reversal, perhaps indicating the use of improved exploration methods such as geophysical surveying.

Figure 2.12 shows the shares of oil and gas discoveries in productive wells by excluding dry wells. This transformation of the substitution process emphasizes the different historical trends in successful oil and gas wells. The "market penetration" of gas wells does not exceed the 10% market share until the 1950s, and thereafter shows a steady increase. It is curious, if perhaps only coincidental, that the beginning of this period of more significant increases in the shares of successful gas wells starts during the 1950s, the period when the largest improvements in drilling depths

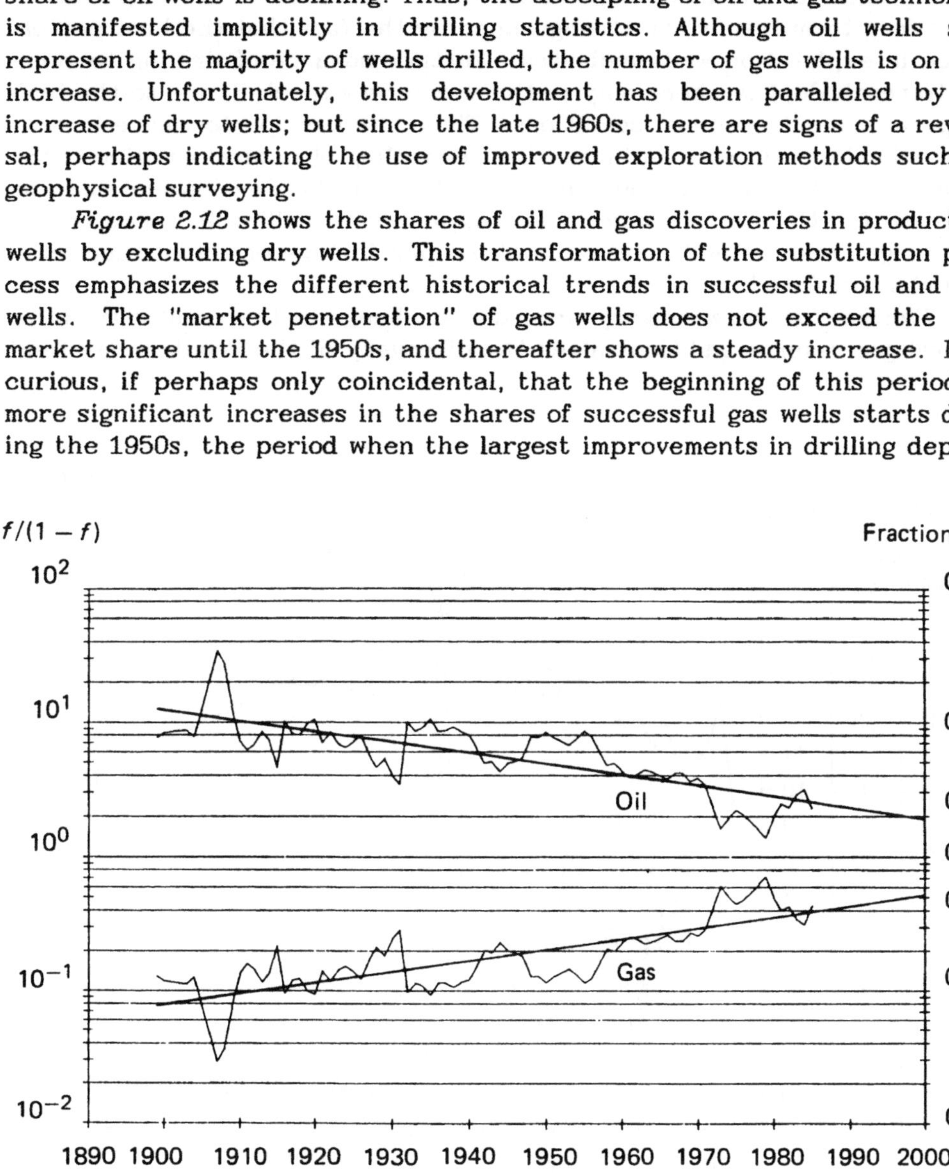

Figure 2.12. US share of successful oil and gas wells.

were achieved, as was illustrated in *Figure 2.8* (the inflection point occurred in 1951). Based on an extrapolation of the long-term historic trend, one might expect for the year 2000 that slightly over 30% of all successful wells drilled in the USA will yield gas and less than 70% will yield oil, compared to a 10: 90% relationship between gas-to-oil finds characteristic of the earlier period.

There is no doubt that the historical trends indicate success ratio improvements in methane exploration and decreases in oil exploration. These long-term trends are strengthened by the fact that the best performance in exploration wells indicates a further advance of drilling technology to reach depths where only methane can be expected, due to the high temperature and pressure conditions. Thus, the best performance in drilling technology suggests an increasing decoupling of natural gas finds from those of oil. An important question is whether similar developments can be observed for the performance of the exploratory drilling in terms of the natural gas reserve additions per well or per footage of actual successful gas wells completed.

Figure 2.13 shows a very pronounced change in the secular trend of natural gas reserve additions after the discovery of the Prudhoe Bay field

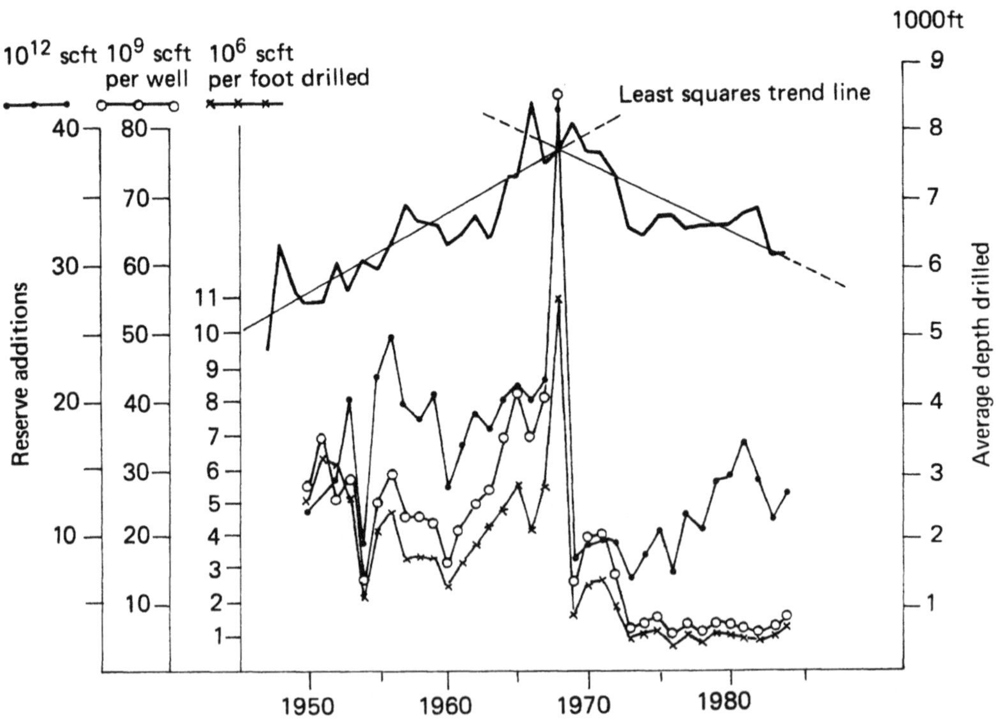

Figure 2.13. US natural gas reserve additions and drilling depth.

in 1969 (the actual reserve additions are backdated to year of discovery). For the figure, the net natural gas reserve additions (from discoveries in new fields and pools as well as the extensions and revisions from known reservoirs) are compared to the number of successful exploratory gas wells (i.e., excluding development wells). Sources of data are AGA (1972, 1975–1979, and 1984) and API (1986). Ever since 1969, the yield per well or per footage drilled has been considerably lower than during the 1950s and 1960s, as shown in the figure. Thus, although the total reserve additions started increasing again after 1969, the additions per well or footage drilled have been rather constant during the last decade despite increases in the share of successful gas wells in the total number of wells drilled (from *Figure 2.5*).

It is perhaps not coincidental that this decrease or lack of improvement in exploratory natural gas drilling in the USA is accompanied by a secular trend change in the average depth drilled (see the least squares trend lines in *Figure 3.5*). The average depth drilled per well increased from about 5,000 feet in the early 1950s to about 8,000 feet by 1970. Thereafter, it has been decreasing, and currently it is slightly deeper than 6,000 feet. Considering the fact that the number of operating rigs has also decreased from about 8,000 to less than 2,000, it is not very likely that another trend reversal will occur in the near future.

Although it is generally dangerous to attempt an explanation based on rather sparse information, it is possible that the decrease in the yield of exploratory natural gas drilling may simply be the result of the fact that wells are tending to become shallower or that the statistics on discovery rates are inaccurate, since it is often a problem of definition as to whether a certain well is successful and what its potential yield might be. Thus, we can speculate on two possible explanations for this change in the yield of exploratory drilling. The first one would simply imply that the wells were not deep enough (as reflected in the decrease of the average drilling depth of some 1,500 feet in the period from 1970 to 1985) to discover large deposits. Consequently, although the share of successful gas wells was increasing with respect to oil wells, only comparatively small fields were discovered, resulting in lower yield rates per well or foot drilled. An alternative (perhaps more plausible) explanation would be that the drilling statistics are distorted by specific conditions of the tax system, resulting in an important number of "tax write-off" drillings not aimed at discovery and subsequent production. If this were the case, such wells would distort the data by introducing a bias into the drilling statistics, e.g., wells would be reported as successful even if only insignificant amounts of gas were discovered, thus decreasing the resulting yield ratios. Nevertheless, *Figure 2.13* suggests that the drilling depth is an important variable to be considered for the potential success and yield ratios of a gas well. This is consistent with the indications that the success ratios (in terms of resource *discovery*) of deep wells (below 16,000 feet) have been extremely high in the USA. In terms of resource *recovery*, however, not all of these wells are reported as successful due to technical or economic infeasibility of production (Whitmore, 1986).

This brief account of technological improvements in drilling technology and historical records of US natural gas finding rates indicate that there is an increasing decoupling of oil and methane technologies, although the traditional drilling technologies were practically identical, whether the well was explored for oil or gas. This decoupling is suggested by ever deeper exploratory wells that exceeded the depths at which one could reasonably expect to find oil during the 1950s. Thus, the deepest wells during the last three decades were all drilled for methane exploration. At the same time, the ratio of successful natural gas wells has increased since 1900, while the ratio of successful oil wells has decreased. Therefore, we find strong evidence that future improvements in drilling technologies will favor natural gas discoveries and improve yields, despite the fact that oil is still the dominant form of energy. However, the future improvements in drilling technologies will also have to result in large cost reductions in order for deep gas deposits to become competitive with other fossil resources.

2.5.2. Oil and natural gas production

Already the generic name "oil and gas" indicates that, during most of its century-old history, natural gas or methane was known as a by-product not only of oil exploration, but also of oil production. During the early days, natural gas was essential, but often a nuisance to the oil producer. Methane pressure in an oil and gas deposit served to pump the oil to the surface, but it was a nuisance because the gas that actually reached the surface had to be flared in order to avoid the danger of explosion. Thus, natural gas was in fact extracted together with oil, but was usually wasted. At the same time, city gas (mostly methane) was being produced from coal and oil to supply premium fuel, especially for lighting and domestic uses. It is not surprising, then, that some associated gas was soon used for consumption.

According to Schurr *et al.* (1960), the earliest recorded commercial use of natural gas in the New World dates back to 1821 (at the time when coal was supplying just 1% of primary energy, and fuel wood and draft animals the rest), when it was used as lighting fuel in Fredonia, New York. Natural gas continued to be used sporadically throughout the nineteenth century. The first pipeline was constructed from Murrysville to Pittsburgh (Pennsylvania) in 1883, after the discovery of a large well in 1878. Despite such pioneering projects by the emerging oil and gas industry, methane was generally considered to be a waste product. By 1878, both crude oil and natural gas passed the 1% share in primary energy consumption, but most of the natural gas consumed in the following decades was used in the vicinity of the oil fields.

Although statistics about natural gas disposal are available only since 1935, they indicate that, at that time, only 75% of the gross natural gas production was marketed and around 25% vented or flared. Natural gas waste was probably considerably higher, since not all venting operations in the course of oil drilling may have been recorded. Natural gas waste decreased, particularly in the period after 1940, through increasing uses of natural gas for repressuring. Still, it is interesting to note that it took until 1974 before natural gas wastes were virtually eliminated, being reduced to account for less than 1% of total gross production, as indicated in *Figure 2.14* (data from Schurr *et al.*, 1960, and AGA, 1972, 1975, 1979, and 1984).

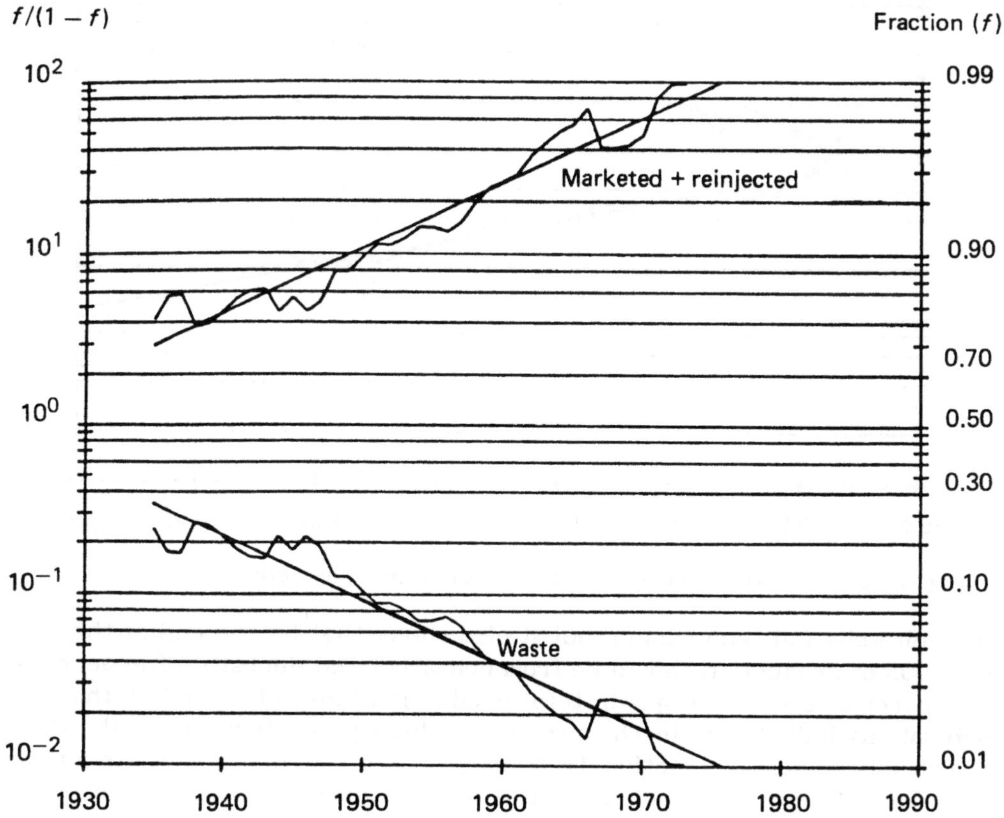

Figure 2.14. US natural gas deposition.

Natural gas production since the turn of the century has grown rapidly, but because, until the last decades, most methane discoveries were related to the search for oil, natural gas extraction was mainly associated with oil production. Accurate statistics of natural gas production from oil and gas wells (associated and nonassociated gas, respectively) are not available for the period before 1935. Hefner (1985) has suggested decoupling associated from nonassociated natural gas production in order to assess the difference, if any, between methane production from oil and gas technologies. *Figure 2.15* shows the steady increase in natural gas production from

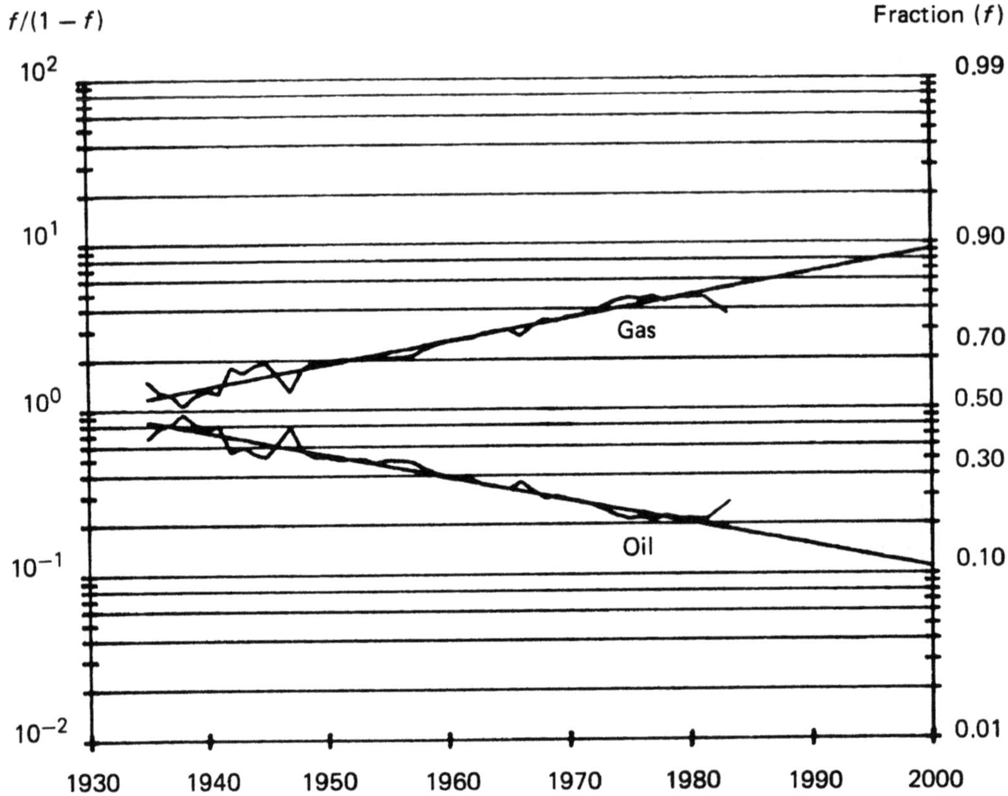

Figure 2.15. US natural gas production from oil and gas wells.

from oil wells (oil technology), but it also shows that by the end of the century 90% of production should be from nonassociated deposits. The decreasing shares of associated gas in total production illustrate the fact that the natural gas industry is in the process of decoupling itself from oil. Thus, the closely related extraction, transport, conversion, and end-use technologies for oil and methane may be slowly diverging toward increasingly independent paths.

This is encouraging, since it indicates a possible next step in the analysis of the evolution of methane technologies. The division of natural gas into oil and gas technologies indicates that it is conceivable that the oil and natural gas industries may also decouple downstream, going perhaps all the way to the final energy consumer. To investigate this possibility, we will next consider the evolution of different energy transport technologies and especially the development of liquid and gaseous energy transport.

2.5.3. Oil and gas transport

During the last few decades, energy became truly a globally traded commodity. Especially crude oil and its products, but to a lesser extent also LNG (liquefied natural gas) and some high-grade coals, are transported around the world. At these global distances, tankers and other vessels are, at least for the time being, the most efficient mode of energy transport. Over continental distances, however, there is a vigorous competition between many alternative transport modes, some of them dedicated to transport of a particular energy form, such as electricity. Crude oil, oil products, and natural gas especially are transported by a number of different transport modes including barges, vessels, trucks, trains, and pipelines.

In contrast to long-distance oil and gas transport to the consumer, wood was primarily consumed locally, close to the source. Some fuel wood was transported over longer distances, mostly by river flotation, and distributed by waterways or roads. Coal, on the other hand, represents a more concentrated form of energy than fuel wood, and coal mines are a more concentrated source than forests (since coal has a higher heat content per unit weight than wood) so that coal was generally transported over longer distances than fuel wood. An extreme case is the modest overseas coal transport (for instance, the coal exports from England to the continent); but more usually coal was transported nationwide by barges, trains, or trucks (earlier by horse wagons). Thus, the shift from a wood- to coal-based economy was accompanied by the expansion of energy transport over longer distances and by an increasing number of transport modes.

The widespread use of crude oil brought another transport mode in addition to tankers, trains, and trucks – oil pipelines. Pipelines are becoming an important freight transport mode with market shares in total ton-kilometers per year comparable to those of train and truck transport. They are also comparable to railways in terms of the total length of the infrastructure or grid: today the total length of main track in the USA is about 200,000 miles, slightly shorter than crude oil pipelines, which total about 230,000 miles. It is interesting to note that in terms of ton-kilometers, car loads, and revenue, coal transport represents by far the largest commodity group in rail transport. Thus, although the total length of the rail and oil pipeline grids is equivalent, the big difference between the two infrastructures is that, since the 1920s, the railroad system has been declining while oil pipelines have been expanding (see Nakicenovic, 1986). *Figure 2.16* shows the rapid increase in pipeline length for crude oil and petroleum products in the USA.

The expansion of the oil pipeline grid parallels the increase of the crude oil share in total primary energy consumption. Oil reached a 1% share in energy during the 1880s, and at the same time the rapid increase in oil pipeline mileage started and followed exponential trends until the 1930s (the inflection point occurred in 1937). As if by coincidence, oil's largest competitor, coal, reached its maximum share in primary energy during the same decade. Thus, by 1937, about half of the current length of the oil

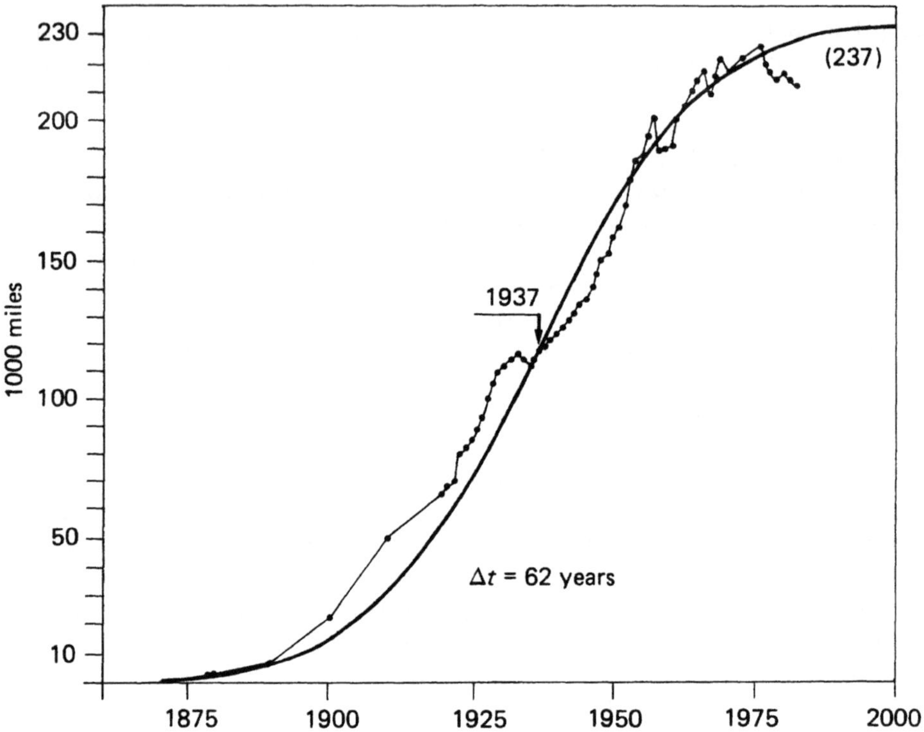

Figure 2.16. US crude and product oil pipeline length.

pipeline network was already in place, and the growth rates declined slowly. During the 1980s, the length should reach the asymptotic level at about the same time as crude oil shares in total primary energy reach saturation. The time constant (Δt) of the expansion of oil pipelines is 62 years or halfway between the time constants for the expansion of rail tracks and surfaced roads of about 50 and 74 years, respectively (see Nakicenovic, 1986).

Although oil is still the most important energy source, in terms of primary energy consumption it is slowly being replaced by natural gas. *Figure 2.15* has shown that the amount of associated natural gas is decreasing in total natural gas production and, therefore, that higher natural gas transport and end-use are based more on gas and less on oil technologies. This process is also reflected in the increase of natural gas transport and distribution pipelines when compared with oil pipelines, shown in *Figure 2.16*. As mentioned above, the first natural gas pipeline in the USA dates back to the 1880s. The rapid expansion of the natural gas pipeline network, however, started during the 1890s, or about 20 years after the growth of oil pipelines was initiated. *Figure 2.17* shows that this 20-year shift in time persists through most of the growth cycle of the natural gas transport system and distribution infrastructure.

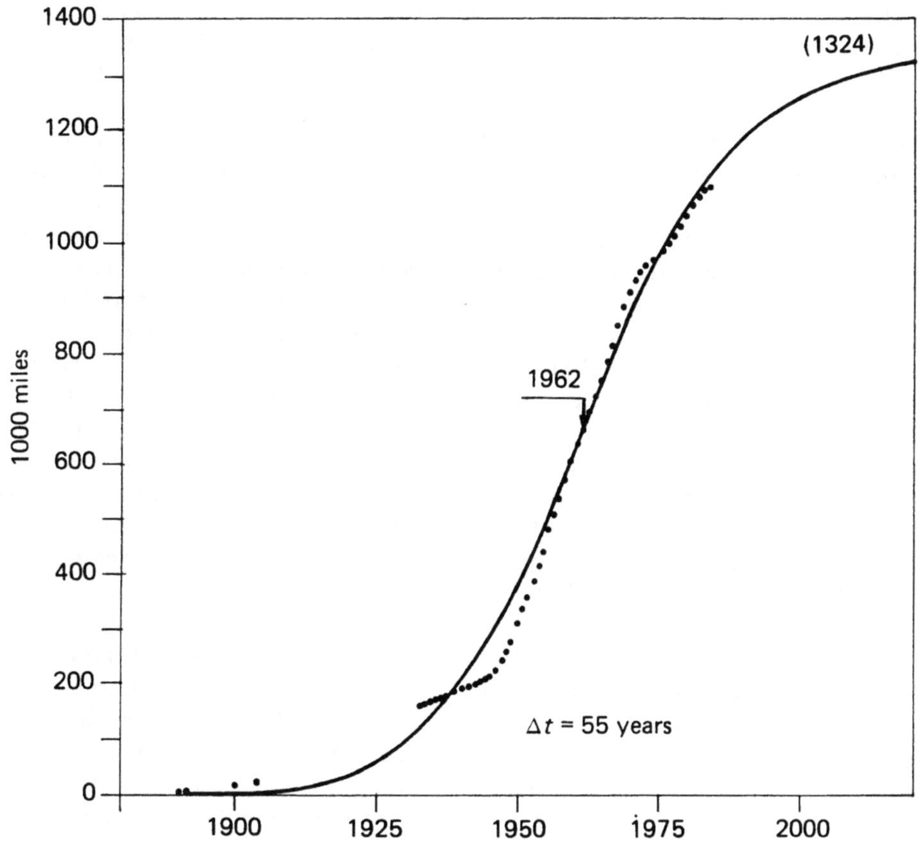

Figure 2.17. **US natural gas pipeline length.**

The inflection point, with about half the eventual saturation level achieved, occurred in 1962, or 25 years after the inflection in the growth of oil pipelines. Since the time constant is about 55 years, and therefore comparable to that of oil pipelines (62 years), saturation should also occur more than 20 years later, during the 2020s. Again, this is symmetrical with the relationship between the growth phases of oil pipelines and oil penetration in primary energy. The growth pulse started when oil achieved a 1% share of primary energy, inflection occurred during the time when oil became the second largest energy source (bypassing fuel wood), and saturation of pipeline length was synchronous with the saturation of market shares. Exactly the same pattern can be observed during the growth pulse of gas pipelines by comparing *Figures 2.17* and *2.3*. The growth started toward the end of the last century when natural gas achieved a 1% share in primary energy. The inflection point was reached in 1962, when natural gas became the second largest energy source (bypassing coal), and saturation of both natural gas market shares and length of pipeline should be achieved during the 2020s.

A large difference between the growth pulses of oil and gas pipelines is in the length of the respective transport and distribution networks. *Figure 2.16* gives a saturation level estimate for oil pipeline length of about 240,000 miles (or about the current length of railroad tracks; see Nakicenovic, 1986), whereas the asymptotic level for the length of gas pipelines is estimated at more than 1,300,000 miles (more than five times higher). For the time being, natural gas is transported almost exclusively through the pipeline grid. Oil and petroleum products, however, are also shipped by tankers, trains, trucks and, for some military use, even by aircraft. Aside from some smaller quantities of liquefied natural gas and liquid natural gas products, most natural gas reaches the consumer either in a gaseous form or as electricity. The pipeline network for gas transport and distribution is therefore also much longer than that for crude oil and petroleum products. This poses the question of whether we can expect natural gas to continue to be transported almost exclusively by pipelines in the future, especially if its projected use expands as dramatically as illustrated in *Figure 2.3*. For liquid natural gas products especially, it is likely that other transport modes will also be used, conceivably even aircraft. From the technical point of view, there are in principle no obstacles to using this transport mode for energy; the only question is whether it would be economical and competitive to do so.

This point cannot be resolved here, but we mention this alternative for the future because similar solutions have been found in the past to meet the ever increasing need to transport more energy over longer distances. Denser and cleaner energy forms were technological measures needed to improve the performance of the whole energy system. We can therefore expect further improvements in the near future, and these could be fulfilled by a stronger reliance on natural gas.

2.5.4 Energy substitution and end-use

Natural gas exploration, production, and transport indicate significantly different trends from oil technologies, although natural gas has been associated with the oil industry ever since its first commercial use. Nevertheless, most energy accounts bind natural gas to oil because of the large production of associated natural gas from oil wells. Except at the point of production, associated natural gas, or oil-technology gas, is indistinguishable from gas produced from natural gas wells. The fact that this distinction is difficult to make, and is consequently ignored in historical data, is to an extent misleading since we have shown that oil and gas technology have followed distinctly different trends during the last century. In addition, the distinction between associated gas and crude oil in terms of primary energy accounting is consistent with adding city gas produced from oil or coal to these primary energy sources rather than to natural gas. To investigate the hypothesis that natural gas is becoming increasingly decoupled from oil technologies, we have attempted to reconstruct the primary energy balances by adding associated gas to crude oil and subtracting the same amount from natural gas consumption (but leaving net imports with natural gas balances).

Figure 2.18 shows the resulting refined version of the primary energy substitution dynamics from *Figure 2.6* in the case of the USA. This revised technological substitution process can be characterized by very regular time constants because the historical data are apparently accurate enough to provide the information required for further analytical resolution. This is possible in the case of primary energy consumption because different energy sources can be measured in common (physical, energy) units and because their use is relatively well documented in the USA for the last 190

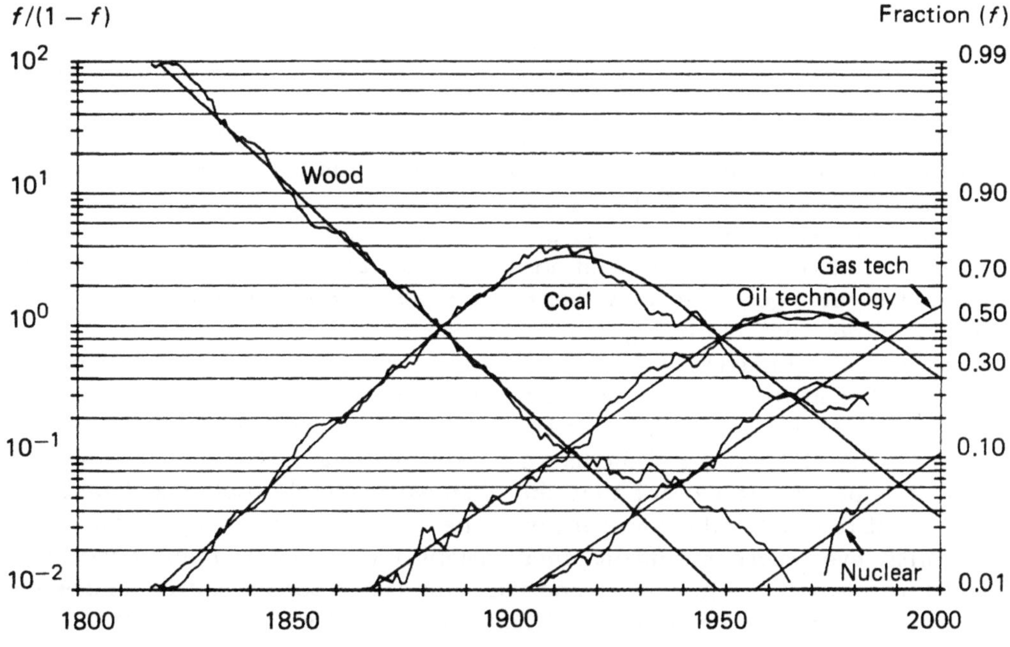

Figure 2.18. US energy substitution and gas technologies.

years. *Figure 2.18* represents such a higher resolution because associated natural gas (oil technology from *Figure 2.15*) is now reallocated to the use of crude oil, leaving only the nonassociated gas from gas wells (including natural gas imports) to "gas technology".

This result shows that, although associated gas has long been available as a by-product of oil, its use does not represent the actual evolution of gas technologies. *Figure 2.18* shows that this refinement of the substitution process improves the regularity to the extent that the time constants now cluster at about 70 years for all energy sources and that the saturation intervals between coal, oil, and gas technologies are all separated by about 50 years. During the saturation periods of the dominant energy sources, new ones are introduced. Gas technologies are introduced during the saturation of coal, and nuclear energy during the saturation of oil. The actual oil shares (with associated gas) have now increased so that the importance of oil during the last decades can be seen as clearly dominating more than half of all energy supplies.

This result is encouraging since it indicates a possible next step in the analysis of the evolution of methane technologies. The division of natural gas into oil and gas technologies suggests that the two industries may also decouple downstream, perhaps all the way to the final energy consumer. In order to investigate this possibility, we consider the evolution of natural gas end-use in different sectors of the economy.

The consumer is faced with the decision of choosing the final energy form of preference and seldom attends to the origin of alternative fuels. Thus, the difference between associated and nonassociated natural gas cannot be made at the level of energy end-use. Instead, we can view the trends in total natural gas use at the level of the whole economy, but cannot distinguish between oil and gas technologies.

Figure 2.19 shows the competition of different branches of the US economy for natural gas. The substitution process is very regular with modest fluctuations of actual data from model estimates [2]. Therefore, the example illustrates a continuous and steady trend toward greater use of natural gas in the residential and commercial sectors. At the same time, the share of natural gas used for industrial purposes is decreasing. This result clearly shows that natural gas does indeed have the characteristics of a premium fuel. Industry usually has more technological opportunities to switch between fuels and to implement environmental control when dirtier fuels are utilized. Residential and commercial sectors, on the other hand, usually represent smaller-scale users in relatively densely populated areas. Thus, it is natural to expect that the cleanest available fuel should be utilized. Next to electricity, natural gas is the cleanest final energy form.

With ever more stringent environmental controls and an increasing need for convenience in energy end-use, this trend toward larger shares of natural gas used in residential and commercial sectors should be expected to continue, as *Figure 2.19* suggests. Considering that many studies indicate that the current economic structural changes in the industrialized countries favor expansion of the service sector compared with manufacturing, *Figure 2.19* implies that this restructuring process would also favor an increased use of natural gas, since increasing shares of natural gas are consumed in the residential and commercial sectors. Thus, growing sectors can be expected to become the largest natural gas consumers, while industry and especially some transport modes may rely more on liquid energy carriers. This does not preclude the possibility that, in addition to electricity, liquid energy forms may be produced from natural gas in the next decades.

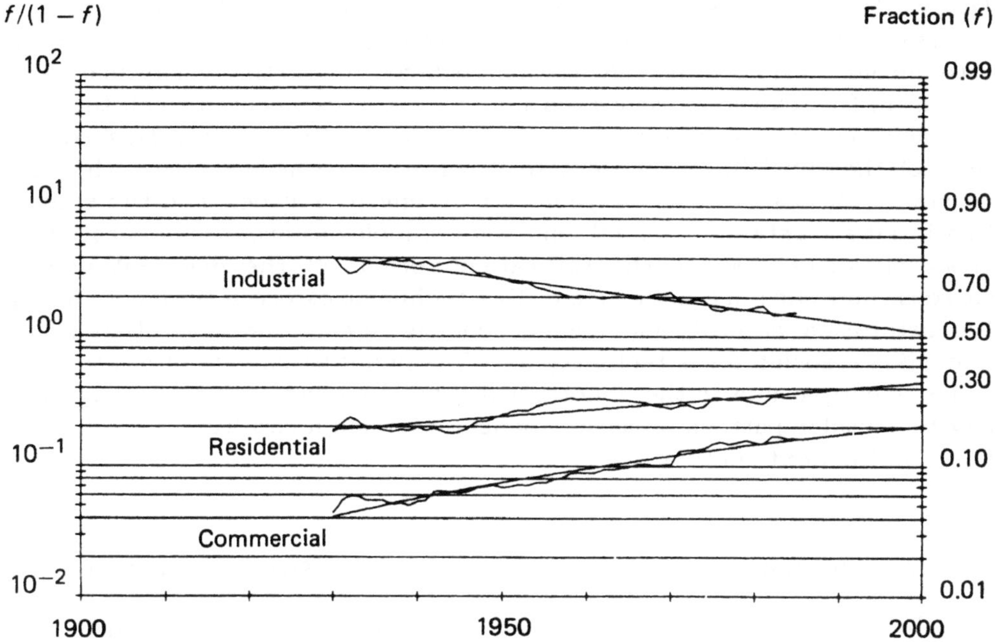

Figure 2.19. US natural gas use by economic sector.

2.5. Conclusions

In a number of examples analyzed in this chapter, we have shown that the growth and senescence of energy technologies can be described as a regular process with logistic secular trends. The dynamics of primary energy substitution in the world and most of the countries and regions analyzed to date (see, e.g., Marchetti and Nakicenovic, 1979, and Nakicenovic, 1984) indicate that natural gas could become the energy source of choice during the coming decades. In this chapter, we have considered this transition from oil to gas as the dominant source of energy at different stages of the energy system from exploration and production to energy consumption and end-use. Throughout the energy system, there are clear indications that natural gas technologies (e.g., exploration and transport) are still in the growth phase, whereas oil technologies are apparently close to saturation. Most of the examples analyzed here are from the USA, primarily because of the US leadership in natural gas use and technologies in the past, and because of data availability.

The refined version of the primary energy substitution process in the USA, based on decoupling oil and gas production technologies as shown in

Figure 2.18, indicates that technological substitution can be characterized by very regular time constants, provided that the historical data are accurate enough to supply the information required for further analytical resolution. In this example, associated natural gas is reallocated to oil technology, and nonassociated gas from gas wells to gas technology. This refinement of the substitution process so improves the regularity that the time constants cluster at about 70 years for all energy sources; and the saturation intervals between coal, oil, and gas technologies are also separated by about 50 years. During the saturation periods of the dominant energy sources, new ones are introduced. Gas technologies are introduced during the saturation of coal, and nuclear energy during the saturation of oil (including associated gas technologies). The penetration rate of gas technologies is comparable to the penetration rate of oil and coal, leading eventually to a similar pattern of market penetration and a resulting market dominance in the next century.

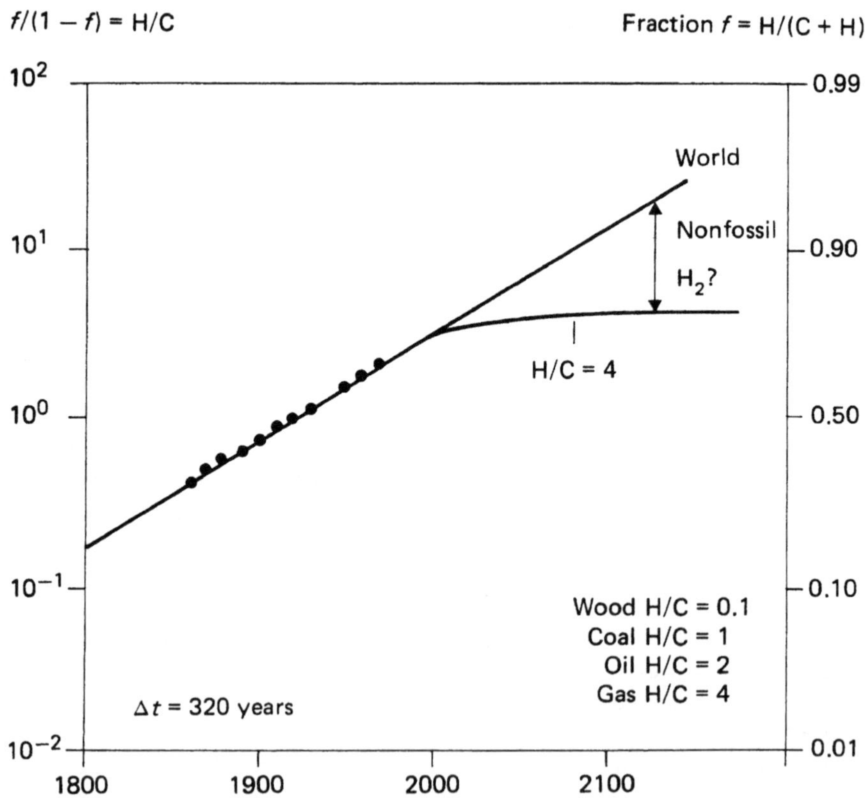

Figure 2.20. World hydrogen-to-carbon ratio of primary energy.

Thus, despite the fact that natural gas use has not increased much during the last few years, in contrast to the growth rates experienced during the earlier decades, our results indicate that the growth potential of natural gas technologies is still very large, especially when compared with oil technologies, which appear to be reaching saturation in the USA. It is very likely that this potential for improving natural gas technology will continue during the coming decades, leading to the development of specific technologies tailored to natural gas that will not be mere derivatives of oil technologies, as they have been in the past. A prerequisite for the widespread use of natural gas in the future is, in our opinion, the decoupling of methane from oil technologies, and this process was initiated long ago when associated gas supplied most of the natural gas consumed. Our results show that the dynamics of energy substitution and methane technologies represent evolutionary, rather than revolutionary, processes that could meet more stringent future requirements through refinements and improvements in current designs and practices during the next two decades. Potentially, methane is cleaner, and it may be distributed abundantly and more evenly throughout the world, than other fossil energy sources.

Thus, methane fulfills most of the obvious future requirements for becoming the major source of energy. A bonus in terms of the very long-term prospects could be that the natural gas economy can pave the way for a very clean hydrogen future. *Figure 2.20* (Marchetti, 1982) illustrates the possibility of such an evolutionary transition in the world from original reliance on fuels with relatively low to fuels with higher hydrogen content; hydrogen-to-carbon ratios were calculated on the basis of actual energy consumption, as specified in *Figure 2.20*, for wood, coal, oil, and natural gas.

This transition is presented as a "substitution" process of hydrogen for carbon in the total primary energy consumption during the last century. *Figure 2.20* shows that, after this century, the hydrogen-to-carbon ratio may exceed the level that can be achieved by pure methane economy, implying that some additional hydrogen could be needed to supplement the increasing reliance on methane. In the meantime, the natural gas share in total primary energy should continue to grow at the expense of dirtier energy sources – coal and oil.

Notes

[1] One general finding of a large number of studies is that substitution of an old technology by a new one, expressed in fractional terms, follows characteristic S-shaped curves. Fisher and Pry (1971) formulated a simple but powerful model of technological substitution by postulating that the replacement of an old by a new technology proceeds along the logistic growth curve:

$$\frac{f}{1-f} = \exp(\alpha t + \beta)$$

where t is the independent variable usually representing some unit of time, α and β are constants, f is the fractional market share of the new competitor, while $1-f$ is that of the old one.

In dealing with more than two competing technologies, we must generalize the Fisher and Pry model, since in such cases logistic substitution cannot be preserved in all phases of the substitution process. Every competitor undergoes three distinct substitution phases: growth, saturation, and decline. This is illustrated by the substitution path of coal, which curves through a maximum from increasing to declining market shares (see *Figure 3.2*). In the model of the substitution process, we assume that only one competitor is in the saturation phase at any given time, that declining technologies fade away steadily at logistic rates, and that new competitors enter the market and grow at logistic rates. As a result, the saturating technology is left with the residual market shares (i.e., the difference between 1 and the sum of fractional market shares of all other competitors – growing or declining) and is forced to follow a nonlogistic path that joins its period of growth to its subsequent period of decline. After the current saturating competitor has reached a logistic rate of decline, the next oldest competitor enters its saturation phase, and the process is repeated until all but the most recent competitors are in decline. A more comprehensive description of the model and the assumptions is given in Nakicenovic (1979).

[2] In fact, this example illustrates the predictive power of the logistic substitution approach in forecasting market shares. This particular example was reported in Marchetti and Nakicenovic (1979), with data up to 1976, so that the period from 1976 to 1985 represents an actual forecast of nine years. Comparison with the original publication shows that the forecast was excellent, within a few percentage points for industrial and residential sectors and precision better than 1% for the commercial sector.

References

American Gas Association (1972, 1975–1979, and 1984), *Gas Facts* (AGA, Arlington, VA).

American Petroleum Institute (1971), *Petroleum Facts and Figures* (API, Washington, DC).

American Petroleum Institute (1986), *Basic Petroleum Data Book: Petroleum Industry Statistics*, Vol. VI, No. 2, May (API, Washington DC).

Brantly, J.E. (1971), *History of Oil Well Drilling* (Gulf Publishing Company, Houston, TX).

Donat, G. (1984), Gaz non conventionnel et gaz profond, *Revue de l'Energie*, No. 366, August–September.

Fisher, J.C. (1974), *Energy Crises in Perspective* (John Wiley and Sons, New York).

Fisher, J.C. and Pry, R.H. (1971), A simple substitution model of technological change, *Technological Forecasting and Social Change*, **3**, 75–88.

Gaz de France (1970), Si le gaz naturel m'était conté, *GdF Informations*, No. 250, January (GdF, Paris).

Gaz de France (1971), Si le gaz naturel m'était conté (2ème partie), *GdF Informations*, No. 262, January (GdF, Paris).

Gold, T. (1985), The origin of natural gas and petroleum, and the prognosis for future supplies, in J. M. Hollander, H. Brooks, and D. Sternlight (eds.), *Annual Review of Energy* (Annual Reviews Inc., Palo Alto, CA).

Hefner, R., III (1986), President, GHK Companies, Personal Communication.

International Petroleum Encyclopedia (1983) (PennWell Publishing Co., Tulsa, OK).

Klemme, H.D. (1977), One-fifth of reserves lie offshore, *Oil and Gas Journal*, **75** (35), August.

Marchetti, C. (1979), Energy systems – The broader context, *Technological Forecasting and Social Change*, **14**, 191–203.

Marchetti, C. (1982), *When Will Hydrogen Come?* Working Paper WP-82-123 (International Institute for Applied Systems Analysis, Laxenburg, Austria).

Marchetti, C. (1986), Fifty-year pulsation in human affairs, analysis of some physical indicators, *Futures*, June, 376–388.

Marchetti, C. and Nakicenovic, N. (1979), *The Dynamics of Energy Systems and the Logistic Substitution Model*, Research Report RR-79-13 (International Institute for Applied Systems Analysis, Laxenburg, Austria).

Nakicenovic, N. (1979), *Software Package for the Logistic Substitution Model*, Research Report RR-79-12 (International Institute for Applied Systems Analysis, Laxenburg, Austria).

Nakicenovic, N. (1984), *Growth to Limits, Long Waves and the Dynamics of Technology*, University of Vienna.

Nakicenovic, N. (1986a), Transportation and energy systems in the US, NAE Workshop on the Evolution of Infrastructures, Woods Hole, MA.

Nakicenovic, N. (1986b), The automobile road to technological change: Diffusion of the automobile as a process of technological substitution, *Technological Forecasting and Social Change* **29**, 309–340.

Offshore (1976–1984), Deepwater well survey, June issues.

Oil and Gas Journal (1977), Petroleum 2000, *Oil and Gas Journal* **75** (35), August.

Peebles, M.W.H. (1980), *Evolution of the Gas Industry* (Macmillan Press, London).

Putnam, P.C. (1953), *Energy in the Future* (D. Van Nostrand Co., New York).

Reynolds, R. and Pierson, A. A. (1942), *Fuel wood used in the United States, 1630–1930*, Circular No. 641 (USDA Forest Service, Washington, DC).

Schurr, S. and Netschert, B. (1960), *Energy in the American Economy, 1850–1975, An Economic Study of Its History and Prospects* (Published for Resources for the Future, Inc. by Johns Hopkins Press, Baltimore, MD).

Taylor, G. R. (1962), *The Transportation Revolution, 1815-1860*, Vol. IV *The Economic History of the USA* (Holt, Rinehart and Winston, New York).

US Department of Agriculture (1946), *A Reappraisal of the Forest Situation: Potential Requirements for Timber Products in the United States*, Forest Service Report No. 2 (USDA Forest Service, Washington, DC).

US Department of Commerce (1970), *Historical Statistics of the United States: Colonial Times to 1970*, Vols. I and II (Bureau of the Census, Washington, DC).

US Department of Commerce (1975, 1980, 1981, 1983, 1984), *Statistical Abstract of the United States* (Bureau of the Census, Washington, DC).

US Department of Energy (1980), *State Energy Data Report, Statistical Tables and Technical Documentation, 1960 Through 1978* (National Technical Information Services, PB80-191224, Washington, DC).

US Department of Energy (1982), *Monthly Energy Review*, August [Energy Information Administration, DOE/EIA-0035(82/08), Washington, DC].

Westfall, L. (1986), Editor, *Ocean Oil Weekly Report*, personal communication.

Whitmore, R.E. (1986), Chairman of the Board, Alaska Bank Corporation, personal communication.

CHAPTER 3

The Future of Natural Gas: A Darwinian Analysis

Cesare Marchetti

3.1. Product Life Cycle

The future is in large measure a deconvolution of the past, with seeds of novelty carefully selected and planted by the system itself. This far-reaching statement is supported by a large body of evidence we have amassed at the International Institute for Applied Systems Analysis (IIASA) during the last ten years.

The first subject analyzed was, in fact, energy at the global level, and natural gas was prominent in the picture we drew. Just ten years ago, IIASA analysts declared its progress to be inevitable, when the press was all against it. The argument was that natural gas should be saved for future generations, so limited were its supplies.

The consequences of that early IIASA analysis were never fully explicated, and the purpose of this chapter is to do so, using the *product life cycle* concept familiar to economists. A primary energy source is, in fact, a product that substitutes for and is replaceable by other products, so the concept is fully applicable. For comparison, I also analyze coal and oil.

The original primary energy competition is reported in *Figure 3.1*, updated to 1982. For coal, the penetration starts at a market share of about 30% (1860), so the previous share had to be reconstructed by logistic backcasting. *Figures 3.2, 3.3*, and *3.4* show coal, oil, and gas, respectively, in terms of their life cycles from an initial 1% of the world market to their final 1% of the market.

For coal, the cycle spans 300 years; for oil, 200; and for gas, 250. This is not really related to size of resources, but only to the timing of the introduction of the competitor. Oil was introduced somewhat "late", according to the general rules of innovation waves and new primary energies introduction

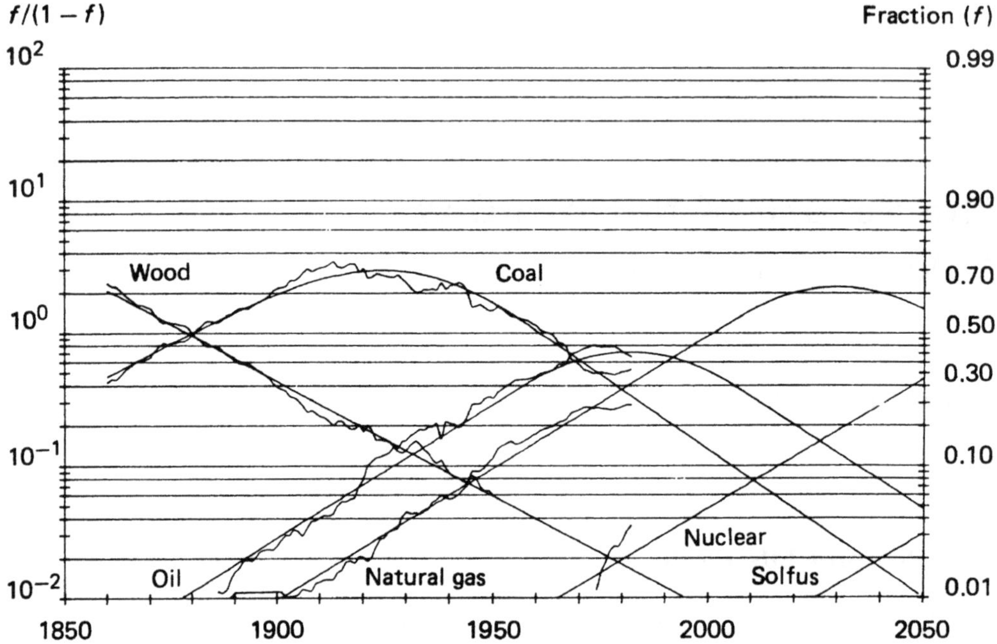

Figure 3.1. World primary energy substitution: market shares (Marchetti and Nakicenovic, 1979; Nakicenovic, 1984). The primary energy market substitution analysis, originally made in 1974, is supplemented with data to 1982. The lower-than-expected penetration of natural gas is a signal for fast growth in the near future. As the past shows, deviations are always reabsorbed elastically.

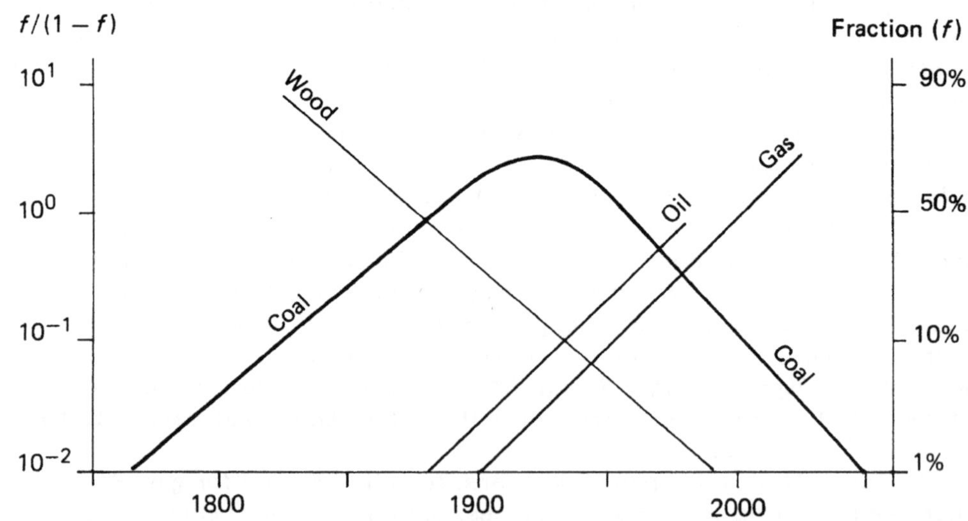

Figure 3.2. World coal life cycle (Marchetti, 1986). Coal's full life cycle has been extrapolated backward for the missing statistics.

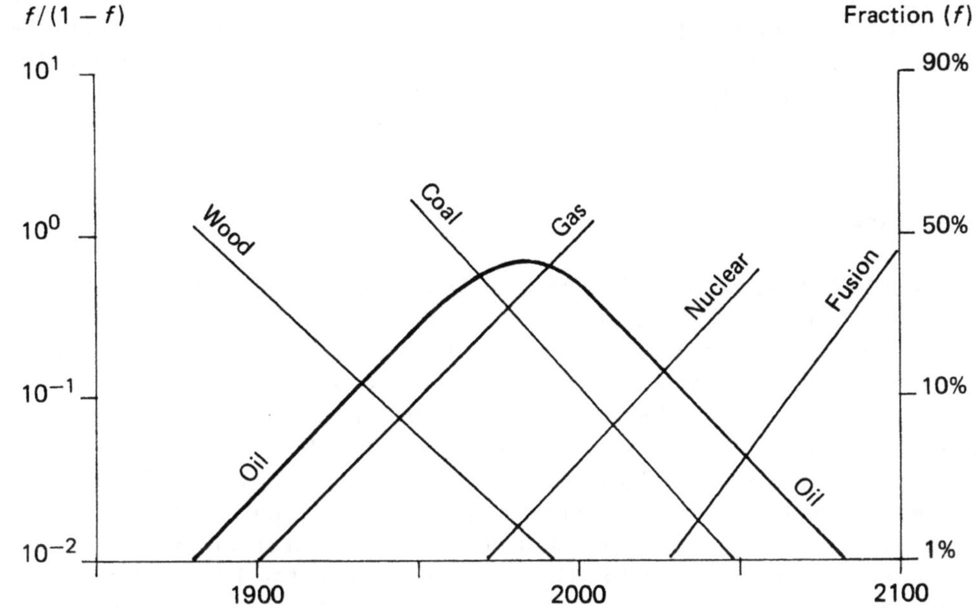

Figure 3.3. World oil life cycle (Marchetti, 1986). The shortness of the cycle is due to the introduction of gas soon after oil.

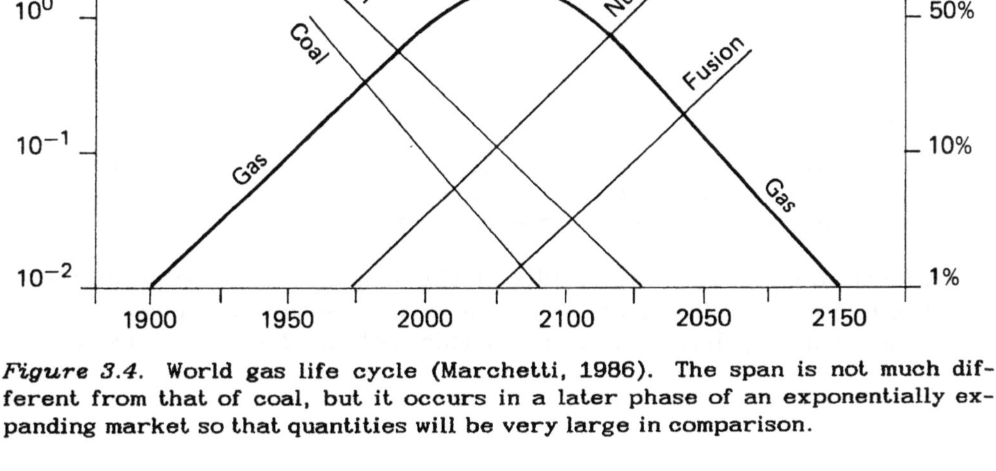

Figure 3.4. World gas life cycle (Marchetti, 1986). The span is not much different from that of coal, but it occurs in a later phase of an exponentially expanding market so that quantities will be very large in comparison.

(Marchetti, 1981). This is the reason for the broad time span of coal and its high maximum market penetration of about 75%. Gas being timely, the loser is oil, with its maximum penetration a little below 50%. Natural gas should go a little above a 70% market share in 2030. But as we will see, because the total market is growing in time, the total amounts extracted during the life cycle increase in the order: coal, oil, and gas.

3.2. Market Growth

In absolute terms, we hypothesize that the world energy market is growing logistically. In the USA, logistic growth begins to show now with a per capita consumption of about 15 tce/year. It seems fair to assume that world consumption, now at about 2 tce/year, will stay exponential for a while (until it reaches the inflection point of the logistic). That has been the case during the last century, with a growth rate of 2.3%/year (i.e., total consumption doubling every 30 years; see *Figure 3.5*). Our time span stretches 150 years ahead and, assuming a doubling of world population every 70 years (at present rates of growth, it would double in 35 years), we land at a world per capita consumption level in the same ballpark as the US per capita consumption *at present*. Thus, our hypothesis about total energy growth during the next 150 years is *business as usual*.

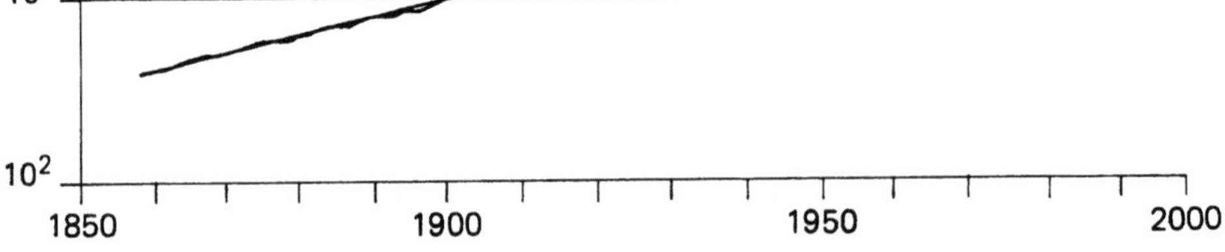

Figure 3.5. World primary energy consumption (Nakicenovic, 1984). Business-as-usual extrapolation of energy consumption at the secular growth rate of 2.3%. The oscillations are due to 55-year Kondratieff cycles.

At this point, I give the results of the simple calculations that follow from the hypothesis. In *Figure 3.6*, world coal consumption is given in absolute terms, showing demand peaking around 1980 at 2.5×10^9 tons. It will

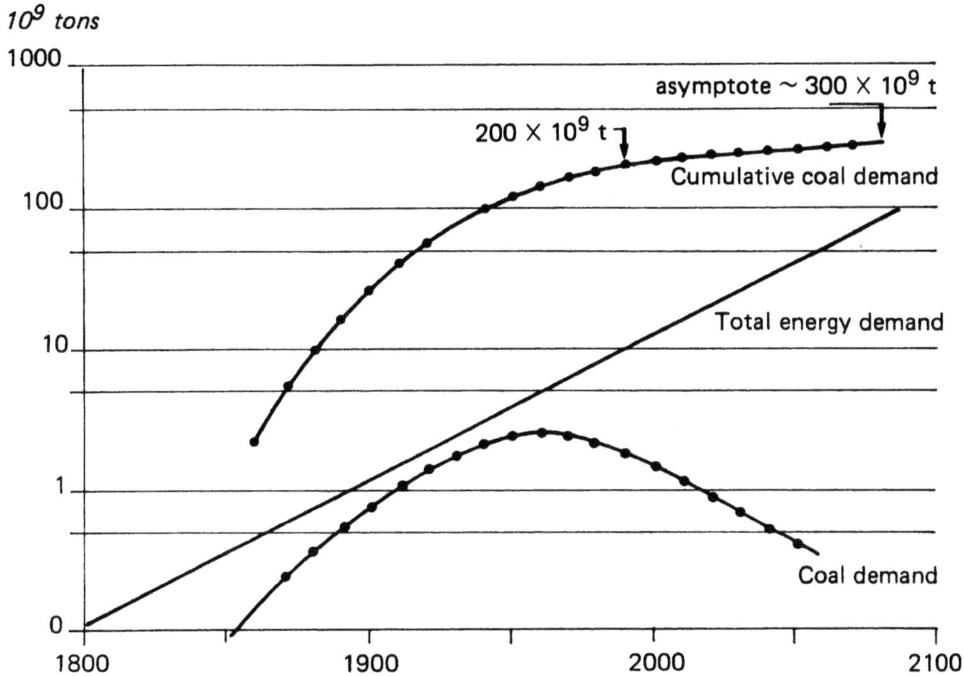

Figure 3.6. World coal consumption with demand in current and cumulative terms (Marchetti, 1986). Total energy demand comes from *Figure 3.5*; market shares, from *Figure 3.1*.

still be about 0.4×10^9 tons in 2050. The decrease in consumption will be well-cushioned, corresponding to a loss of about 2.5% per year. The integrated amount extracted during all of the coal-age cycle will amount to $\sim 300 \times 10^9$ tons, of which $\sim 200 \times 10^9$ tons have already been extracted. As people are now so sensitive to deaths linked to energy use, it would be interesting to calculate how many people may die just to extract that final 100×10^9 tons. My back-of-the-envelope calculation is about 200,000 – taking into account the fact that much of the future coal supply will be mined in developing countries.

Figure 3.7 shows the consumption of oil. Oil demand is at a maximum now and will stay basically the same for another 20 years at a little below 3×10^9 tons. All curves in *Figure 3.7* are smooth because total energy consumption has been smoothed by eliminating the Kondratieff perturbations that clearly appear in *Figure 3.5*. Integrated oil demand will go to an asymptote of 400×10^9 tons, of which only about 100×10^9 tons have been extracted. Although already at the beginning of its decline, the oil industry still has most of its future in front of it. Demand will still be 0.5×10^9 tons in 2100, which corresponds to a *mean* loss of production of 1.6% per year.

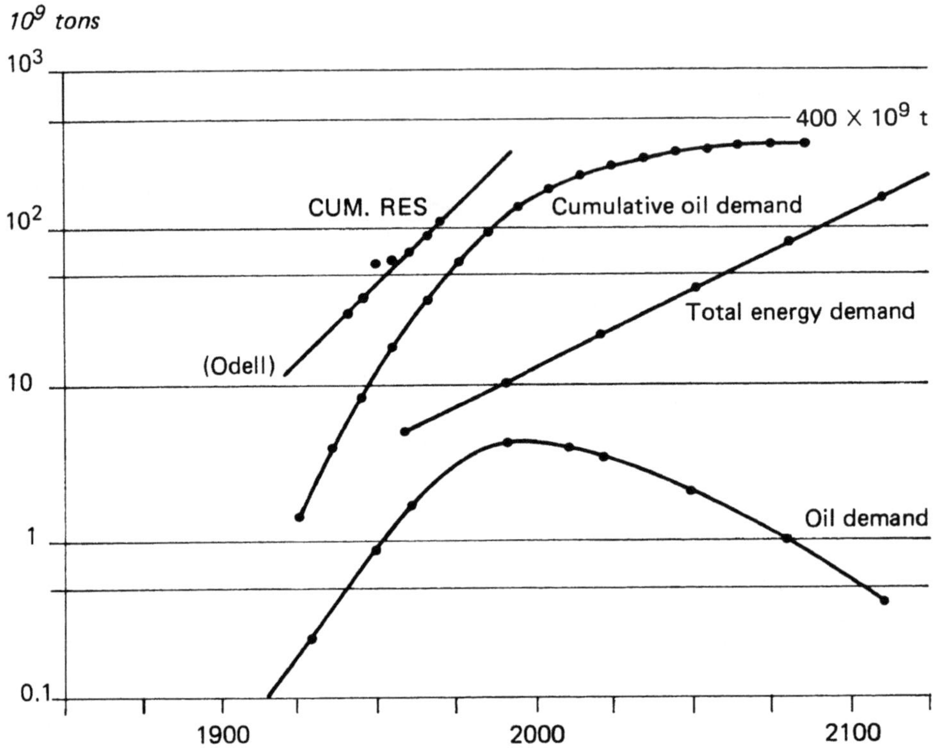

Figure 3.7. World oil consumption in current and cumulative terms (Marchetti, 1986). Total energy demand comes from *Figure 3.5*; market shares, from *Figure 3.1*.

In *Figure 3.8* the consumption of natural gas is reported. All the numbers appear very large because this technology is developing within an expanding market. Gas demand will peak around the decade from 2060 to 2070 at the level of about 30 × $10^{12}m^3$. This is roughly an order of magnitude higher than the peak for oil (in terms of energy content). Gas demand on its way down would reach the present oil demand level around 2150. The asymptotic level for integrated demand is about 2,500 × $10^{12}m^3$, *of which less than 2% has already been extracted.*

On top of the assumption of a stable 2.3% growth rate for total energy, I also assumed that the time constant (Δt) of penetration for nuclear energy is about 80 years – more or less business as usual. The rate of penetration of fusion, or of some later developments, will have no influence because the rate of decrease will be already set up by the overall penetration rate of nuclear energy. This confidence comes from the rules of market substitution. The rates of decrease of coal and oil are already set at present.

Because the strength of all these deductions depends on the validity of the hypotheses, let us look again at them critically. The capacity of a system of interacting logistics to predict the market share over long periods of

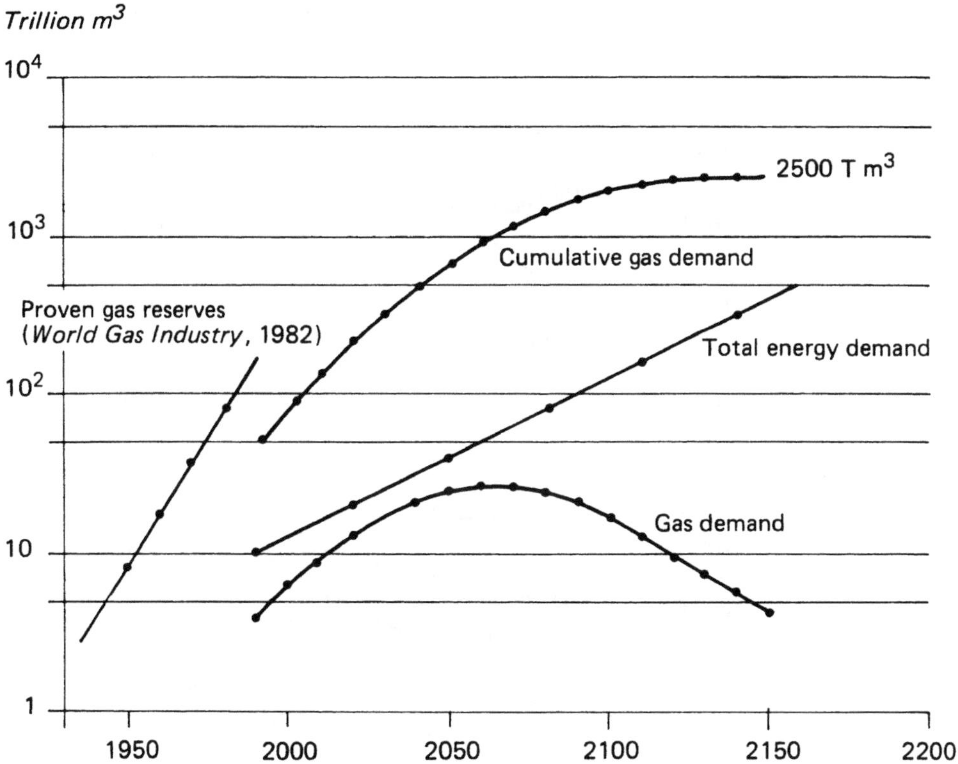

Figure 3.8. World natural gas consumption in current and cumulative terms (Marchetti, 1986). Total energy demand comes from *Figure 3.5*; market shares, from *Figure 3.1*.

time has been tested in many cases. *Figure 3.9* (*a*, *b*, and *c*) reports our classical example, showing that the world market share of oil in 1980 could have been predicted in 1920 within a precision range of a few percentage points.

The growth of a certain commodity (here, energy) to fill the "niche" is always logistic, i.e., almost exponential below the flex. For energy, we are not yet equipped to calculate the saturation point, so the exponentiality may become weaker toward the end of the period analyzed, leading to an overestimate of consumption, e.g., for the 50 years after 2100.

As for population growth, which serves to check more than to prove, demographers are sure of the next doubling to 10 billion, but most specialists have no opinion about the subsequent doubling. I constructed a scenario in which the earth could accommodate 10^{12} people, just to test the size of the niche (Marchetti, 1979). No niche was left half empty during the last four billion years!

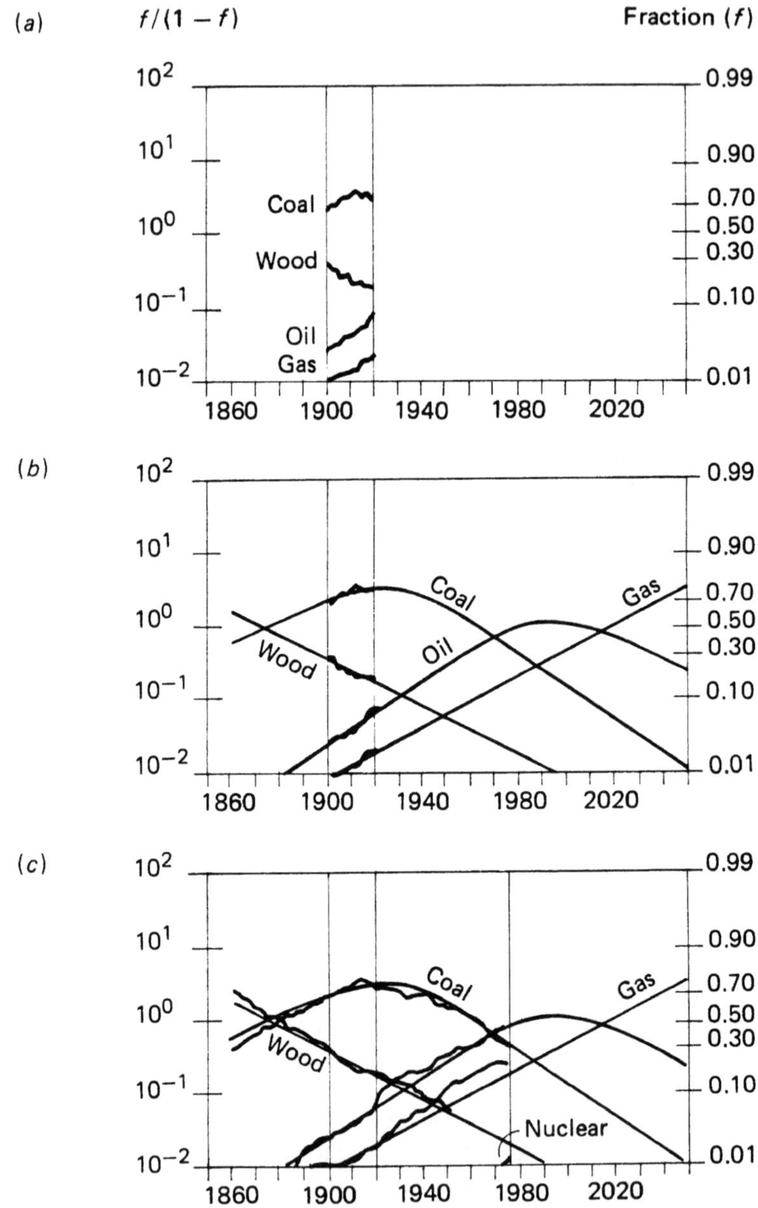

Figure 3.9. World primary energy substitution (short data). (*a*) To show the potential of market substitution analysis for prediction, the case of a limited data base (1900–1920) is reported. (*b*) The equation is established using this data base. (*c*) Actual statistical data are superposed to check the forecasting. Only gas appears relatively poorly predicted, but this is due to its very low penetration (2%) at the end of the data base.

3.3. Resource Base

How do the figures for integrated demand compare with resources and reserves? For coal, there is no problem: 100×10^9 tons are already in known deposits; only logistics and politics affect the search for new ones. For oil, the situation is a little tighter, but 300×10^9 tons seem to be in the ground. As *Figure 3.10* shows, the estimates of final resources glide over time. Because there is no general theory of oil formation that permits an "external" estimate, one has to drill and look. The process being very expensive, consumption controls the search activity. As *Figure 3.11* neatly shows, oil reserves in the USA have *always* equaled about 14 years' worth of current consumption.

In a 1976 publication, B.F. Grossling of the World Bank pointed out some peculiarities of the process, e.g., that the search for hydrocarbons is controlled much more by geopolitical context than by the probability of discovery (*Table 3.1*). That maximum drilling activity occurs where discovery success is minimal defies any viable economic explanation. Furthermore, because the only sure way to know is to drill, the density of drilling over possible hydrocarbon-carrying areas is the best measure of its potential. As reported in *Figure 3.12* and *Table 3.2*, most of the world is still virgin.

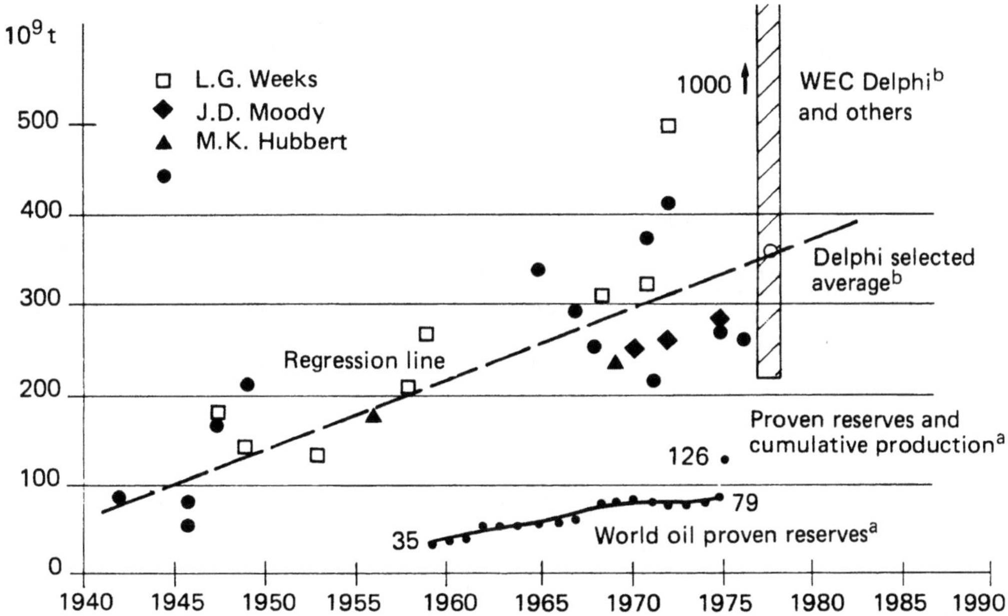

[a] World Oil – August Issues 1960 to 1975.
[b] Conservation Commission. Report on Oil Resources.

Figure 3.10. Best estimates of final oil resources (*World Oil*, 1977). The clear upward trend is more or less in tune with exploration levels

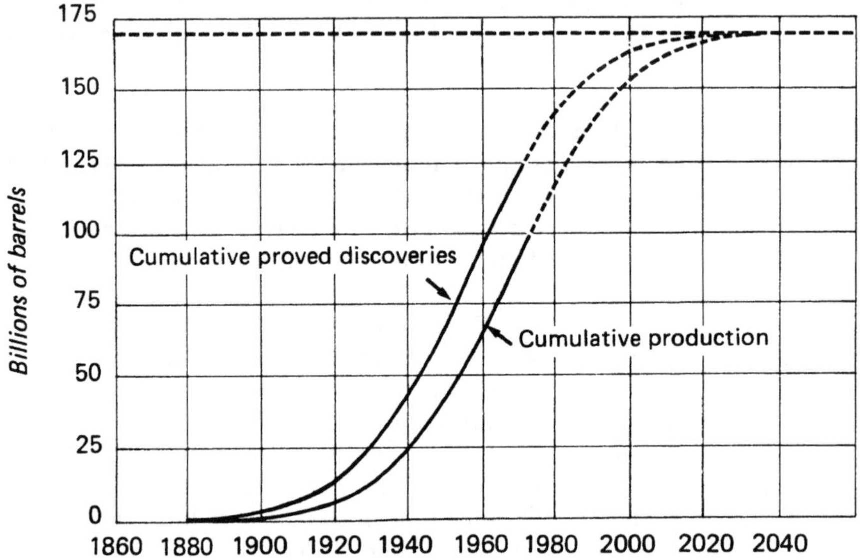

Figure 3.11. Cumulative proven discoveries and production for the USA since the beginning of the oil era (Hubbert, 1976). It is clear that economics strictly links research to production. The gap between the two curves is a very stable 14 years.

Table 3.1. Drilling finding rates (tons per meter of exploratory well).

Year	*USA*	*Western Europe*	*Latin America*	*Africa*
1970–1974	7	520	200	520
1960–1964	6	17	57	380
1950–1954	7	40	80	37

Source: Grossling (1976).

Table 3.2. Free world exploratory well completions (rounded figures).

Region	*1970*	*1972*	*1974*	*1976*
North America	9,000	9,000	10,500	12,000
Latin America	500	400	400	400
Eastern Hemisphere	700	800	800	800
Africa	250	200	200	150
Europe	150	200	250	250

Source: Grossling (1976).

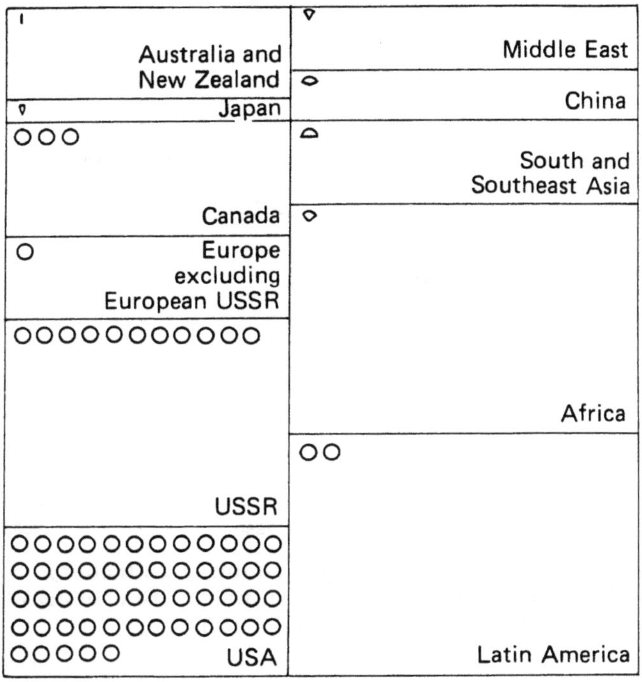

Figure 3.12. World petroliferous regions: perforation level in proportion to land area (Grossling, 1976).

For natural gas, the final integrated demand is five times that for oil. It would be consoling to find some geological hint that this demand can be met. Professor Gold's arguments (Chapter 5 in this volume) are of capital importance, and we pray that the Siljan drilling project will reveal a gas deposit in primary rock at least as large as the Groningen area. We need just 1,000 of such deposits. In the meantime we have to rely on Grossling's (1976) logic: Because known reserves are on the order of $100 \times 10^9 m^3$, and the volume of potential gas-bearing formations already explored may be on the order of a few percent (*Figure 3.12*), it is possible that final resources will match the total forecasted demand.

3.4. Product Delivery

For natural gas, long-range transportation is the weakest point of the present system. *Figure 3.13* traces the development of world gas trunklines, i.e., pipelines with a diameter larger than 12 inches, the long-distance work horses. The saturation point is about one million kilometers near the

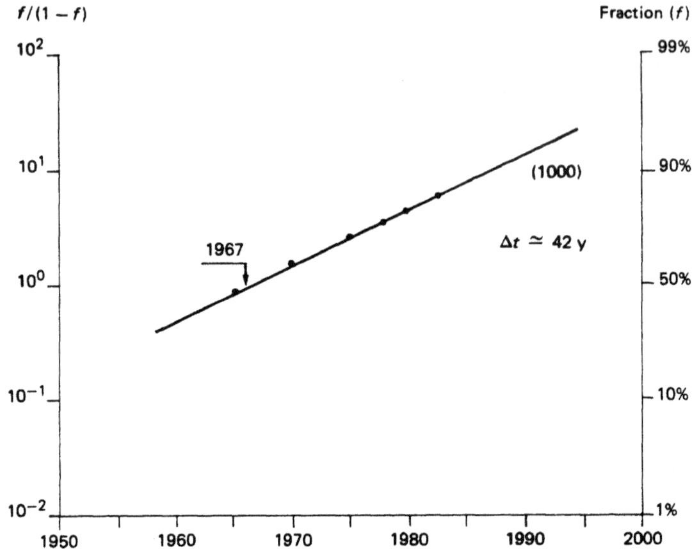

Figure 3.13. World gas trunklines (10^3 km: radius > 12 inches). Saturation point for the present Kondratieff cycle is 10^{10} km, half of that being already in place in 1967. (Source: based on data from Technip, 1982.)

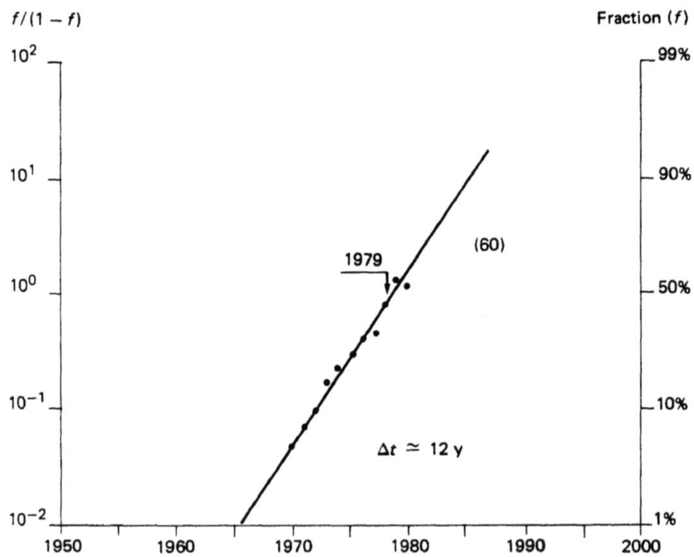

Figure 3.14. World LNG commercial market (10^9 m^3). Saturation will be reached soon at around 60 × 10^9 m^3 of gas. The time constant is very short, as is characteristic of technologies born near the end of a Kondratieff cycle. (Source: based on data from Technip, 1982.)

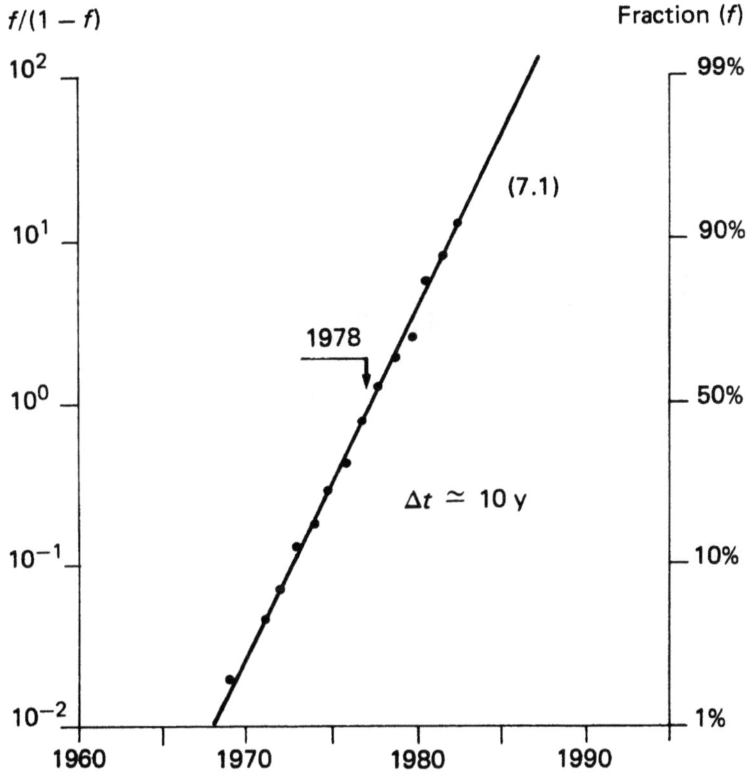

Figure 3.15. World LNG tanker fleet (million tons). The size of the tanker fleet closely mirrors that of the market and saturates at 7.1 million tons. That growth is slightly slower than for the market may signal improved utilization of the ships or shorter routes. (Source: based on data from Technip, 1982.)

end of this century, which does not contradict the previous forecasts, because the Kondratieff oscillations are smoothed. Almost everything "saturates" near the end of a Kondra cycle (1955!), and pipeline networks are no exception. Expansion on a new logistic will start again toward the end of the century. Incidentally, because of the increase in methane consumed (i.e., transported) by more than an order of magnitude during the next Kondra (55 years), the bore of the largest pipes, now 58 inches, should double. This link between traffic and size is very evident for electricity networks, for example.

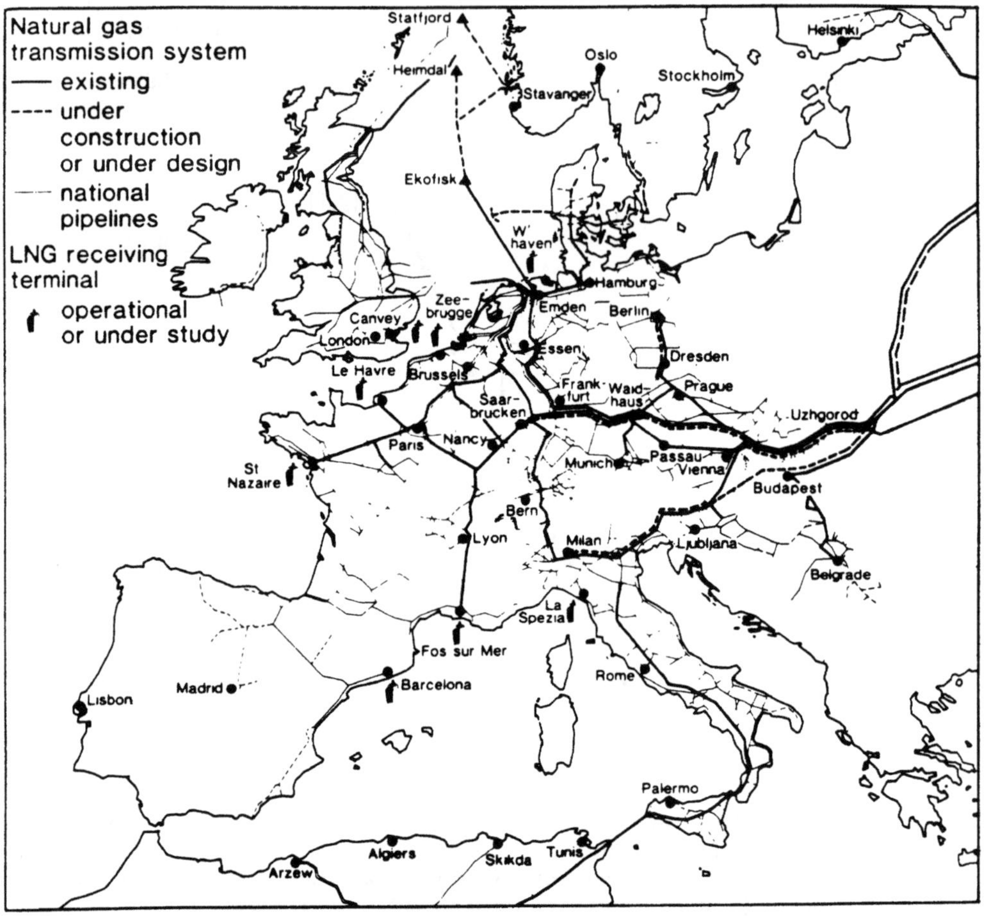

Figure 3.16. European natural gas transmission system. Source: Odell (1985), based on 1984 data provided by Ruhrgas.

Twenty years ago, the first LNG tankers for overseas transportation inspired much enthusiasm. The trade developed nicely, but not explosively. *Figure 3.14* gives LNG transported worldwide, saturating at around 60 × $10^9 m^3$. LNG tanker fleets should follow a similar trend (*Figure 3.15*), which is in fact perfectly matched by a logistic, saturating at 7.1 tons, with a similar time constant and central point. The introduction of the technology was very fast, and it is by now saturated. We must await the next wave (1995) to see if a technical breakthrough or a surge in this type of demand produces a new logistic.

Europe, with its network of gas transport lines (see *Figure 3.16*), has clearly entered the methane age. It is not yet clear whether metropolitan resources will be capable of sustaining all that development. Certainly, the outlook would improve if the Soviet Union were to enter full force into the picture. This possibility is obviously linked to many factors, basically political ones.

A prudent policy would be to structure the whole system so as to be resilient in the event of default by one of the large suppliers. Such a scheme was developed by my collaborators and myself a dozen years ago (Marchetti *et al.*, 1973), based on the strategic use of inboard gas resources. Large reserve production capacity would always be kept alive, and transport lines (e.g., Russian ones) would operate at full capacity. Surpluses would be stored in consumer areas. These procedures are currently in use in order to satisfy winter peak demand. We suggested that they be viewed in a strategic mode, at the geographical level of Europe, and on a time scale of ten years. The critical variable in the European natural gas calculus is likely to be the attitude of the Soviet Union toward schemes of that kind.

Acknowledgment

Earlier versions of this chapter appeared in *Bollettino degli Ingegneri* **34** (September, 1986), © Collegio Ingegneri della Toscana; and *Technological Forecasting and Social Change* **31**, © 1987 Elsevier Science Publishing Co., Inc.

References

Grossling, B.F. (1976), *Window on Oil: A Survey of World Petroleum* (Financial Times, Ltd., London).

Hubbert, M.K. (1969), Energy Resources, in P. Cloud (ed.), *Resources and Man, Committee on Resources and Man of the National of the National Academy of Sciences – National Research Council* (W.H. Freeman, San Francisco, CA).

Marchetti, C. (1979), 10^{12}: A check on the earth-carrying capacity for man, *Energy* **4**, 1107–1117.

Marchetti, C. (1981), *Society as a Learning System: Discovery, Invention and Innovation Cycles Revisited*, Research Report RR-81-29 (International Institute for Applied Systems Analysis, Laxenburg, Austria).

Marchetti, C. and Nakicenovic, N. (1979), *The Dynamics of Energy Systems and the Logistic Substitution Model*, Research Report RR-79-13 (International Institute for Applied Systems Analysis, Laxenburg, Austria).

Marchetti, C., Rinaldi, C., and Schneiders, A. (1973), *Model of a Strategy for the European Energy Supply Based on Methane as the Prime Energy Carrier*, EUR/C-IS/479/73e (Commission of the European Community, EURATOM Joint Research Center, Ispra, Italy).

Nakicenovic, N. (1984), Personal communication.

Odell, P.R. (1985), *Natural Gas in Western Europe: Major Expansion in Prospect*, EURICES Paper No. 84/5 (Centre for International Energy Studies, Erasmus University, Rotterdam).

Technip (1982), *World Gas Industry*.

World Oil (1960–1977), August issues.

CHAPTER 4

Natural Gas and Technical Change: Results of Current Gas Studies

Hans-Holger Rogner

4.1. Introduction: Typical Natural Gas Analyses

At the outset of his contribution to the Sixteenth World Gas Conference, M.W.H. Peebles (1985) presented the sequence of distinct steps conventionally adopted in assessing future supply and demand for energy in general and for natural gas in particular. The principal components of major natural gas studies roughly cover:

(1) The business perspective:
  (a) Industrial activity and structural change.
  (b) Energy conservation, price responsiveness.
  (c) Environment.
  (d) Energy policy.
(2) The outlook for oil: prices versus supplies.
(3) The prospects for natural gas:
  (a) Resource base.
  (b) Interfuel competition.
  (c) Gas-to-gas competition (recently).
  (d) Spot market (recently).

Typically, the assessment of the prospects for natural gas begins with the conceptualization of an overall business perspective. Since the era of predominant dependence on oil is likely to last well into the twenty-first century and economic performance continues to be highly sensitive to the dynamics of the international oil market, the next step defines an outlook for oil. Only then does the analysis turn to the potential role for natural gas within the setting defined by the two preceding steps. However, one essential factor is omitted from most gas studies. *Technology*, though implicitly present the form of static reserve-to-production ratios or recovery rates, seldom enters energy projections as a variable on its own merits. These ratios are one-time snapshots based on existing exploration and production technologies, neglecting the aspects of dynamic technical change. Consequently, these studies disregard the impact of technical progress on the composition of future primary energy supply mixes. The seriousness of this omission is illustrated by comparing the results of several recent gas studies that follow the conventional format.

4.1.1. The business perspective

The overall business perspective serves as the background against which future energy demand and supply patterns are customarily assessed. Basically, this step is an attempt to identify the economic and political climate at the macro level and aims at quantifying the dynamics and trends of those factors impinging directly or indirectly on energy demand. These include, first of all, future economic growth patterns, i.e., the volume and quality of economic activity and the effects of industrial structural change on energy intensities.

Second, energy conservation, which in the past was the result of unexpected consumer response to seemingly ever rising energy prices, is a critical variable, especially in the context of extended periods of low oil market prices. The question of permanent versus temporary conservation effects also becomes an issue.

Third, a host of government regulatory constraints and energy policies affects the level and mix of energy consumption. Environment-oriented regulations toward reduced stack emissions change the economics and preference for individual fuels, though not necessarily in a consistent way. For example, electricity is highly preferable at the point of consumption, which is not always the case at the point of generation. Energy policy influences both the absolute levels of energy demand and supply as well as the composition of energy consumption. Most notably, there is the dichotomy between the perceived value of domestic resources versus the critical level of dependence on foreign supplies, particularly from politically diverse sources. Energy policy usually means interference with, and distortion of, the free interplay of market forces and may result in an unfair advantage or disadvantage for a particular fuel versus its competitors.

4.1.2. The outlook for oil

The outlook for oil, i.e., its availability and international market price, remains the yardstick for all other energy forms. However, in an effective market environment, one would not expect any criterion other than burner-tip competitiveness to be the market clearing mechanism. It is the inherently long lead times required for any significant change in the demand and supply structures that preserve the dominating role of oil and sensitize energy markets.

In the recent past, even minor changes in the oil business have had major destabilizing effects. Consider OPEC's shift from the position of the world's principal supplier to a swing producer. It is the OPEC multiplier phenomenon (Mossavar-Rahmani, 1984) – i.e., the combined effects of investment dynamics within the typical business cycle and the overreaction of virtually all agents involved in the production, transport, refining, distribution, and consumption of oil – which underscores the quasi-reference role of oil in current and future energy markets.

In summary, the potential for interfuel substitution for oil will depend, over and above burnertip competitiveness, on the perceived availability and the anticipated price of oil.

4.1.3. Prospects for natural gas

Factors most relevant to natural gas include an appreciation of the resource base, the interfuel competition, and contract and pricing issues – all of which center around the specific properties of methane. Resources are plentiful, but distances between resource locations and consumption centers have been continuously expanding. This aggravates the traditional transportation disadvantage of natural gas. In addition, geopolitical concerns come into play whenever the large-scale use of the world's largest known gas reserves – those of the Soviet Union and the Middle East – is considered.

Interfuel competition for natural gas is usually restricted to stationary applications, which by definition exclude the transport sector. Furthermore, the decomposition of methane requires high-temperature heat, thus limiting its use in the feedstock market.

The pricing of natural gas, to a large extent, is subject to institutional influences and policies, with free market forces and competitiveness often playing no more than a marginal role.

Because of the recent surplus, gas supply capacity in major gas markets, gas-to-gas competition, or signs of a possible development of a spot market have received attention. Here, the main concern relates to the adverse effects of price-cutting (to recover some of the sunk capital) on the producer's incentive to invest in new production capacity. Also, spot sales have created some unease in an otherwise precisely "contracted" market.

4.2. The Missing Element

The common approach underlying most gas studies, together with the supply uncertainties arising from political events of the 1970s, the wide fluctuations in energy prices, plus major world fuel trade and consumption distortions, have led to very cautious and almost uniform energy demand projections. The modest energy demand growth expectation for the remainder of this century and the present oversupply situation have in many areas alleviated concerns about energy supplies. A flamboyant demand-creates-its-own-supply attitude prevails in many energy-related fields.

During the early 1980s, when the contracted market began to break down and the long-term supply from secure sources became an issue, quite a few gas studies were conducted by, among others, the International Gas Union, the International Energy Agency, the Commission of European Communities (Guilmont *et al.*, 1986), Data Resources Inc. (1984), the Massachusetts Institute of Technology, and Harvard University. By and large, these studies adhered to the principles of economics, e.g., "Natural gas must be priced competitively downstream" (Harvard International Gas Study, 1985) or "The ability of the industry to bring remotely located reserves to market at the right time, in sufficient quantity, and at marketable prices which at the same time offer the suppliers an acceptable economic return" (Peebles, 1985), as the decisive criteria in their analyses. However, economics, to a large extent, derives its real-world predictive power from policy and technology. Both factors were usually neglected in the studies cited above.

Regarding policy, the pervasive influence of governments and their energy policies is often acknowledged, then treated as a fixed parameter over the entire projection horizon. But government policies are subject to change, a fact disregarded by many energy–gas studies. Self-fulfilling prognosis is often the outcome of this procedure. For example, a policy that inhibits gas from being used as a fuel for electricity generation will, if maintained in an economic analysis, exclude gas from the supply mix of the electricity sector despite its potential economic viability. Planning for the future based on such intelligence might forfeit valuable market shares for natural gas.

Even more significant than treating energy policies as constant are the consequences of constant energy technology projections. The major forces of the last decade that countered the "running-out projections" of the Club of Rome, the perception of an inelastic energy demand, or the power of OPEC are, to a large extent, attributable to technology. The term technology as used in this context includes productivity and efficiency increases as well as infrastructural adjustments – in short, all those elements that can be subsumed under "technical progress". It is intriguing that the evolution of technical progress has been ignored in most gas studies and their specific projections and assessments have been based on state-of-the-art technology only.

There are exceptions. For instance, the Gas Research Institute (1985) published a revised technology scenario after recognizing that technological developments within GRI-funded R & D programs conflicted with projections in some critical areas. Another exception to this constant technology scenario can be found in Tempest (1986).

Unlike energy policy, technology cannot change overnight. But invention and innovation, the driving forces behind technical progress and economic development, are always present, though often undetectable in the short run. Hence, in a study concerned with short-term forecasts of natural gas prospects, e.g., gas sales for the next quarter or so, technological change will not be a critical variable. But in any study that looks ten or more years into the future, technology may impinge upon the market realities at the same rate as does energy policy. Although we are unable to predict future policies, one can be sure that tomorrow's technology will differ from today's. A sound measure of understanding of technology dynamics is a prerequisite for any longer-term energy demand and supply planning exercise. Strategic planning depends on a consistent analytical framework that should, at least, account for potential technological change. Historical developments can serve, at least, as zero-order guidelines when specifying assumptions for future technology evolutions.

Over the last 150 years, the global energy system has experienced a number of evolutionary changes as the fuel mix shifted from the dominance of wood to coal and subsequently to oil (see *Figure 2.1*). These substitutions of one form of primary energy for another have occurred in collusion with fundamental technological change and are technology rather than energy substitutions. Furthermore, these shifts have been independent of the actual resource situation, i.e., the world was not running out of coal when oil began to displace coal. Even economics was not the initial driving force until an oil-accommodating infrastructure was developed and the learning curve and economics of scale became effective. The reason for oil's success in increasing its market share against coal is that oil is superior in many ways and better suited to society's needs. A least-cost energy supply analysis performed at the beginning of this century would most likely have suggested that the coal supply would experience a squeeze in a few decades, and would have missed oil's potential entirely. [No doubt, if a cost–benefit analysis had been made at the crucial time, then sail would never have given way to steam (Ravetz, 1986)].

Our situation today resembles the initial stages of oil's substitution for coal. Oil resources are still plentiful, but society has been sensitized by the destabilizing events of the 1970s, recognizing the geopolitical risk of today's dependence on OPEC oil, and has become increasingly aware of the environmental consequences of fossil fuel consumption. Natural gas offers unique opportunities for improvements on all the above counts. Gas efficiencies in end-use conversions exceed those of oil and thus enhance energy conservation. Methane emits less carbon dioxide than oil, and it is essentially free of sulphur dioxide. Moreover, methane is geographically a more evenly distributed resource, even by traditional resource assessment standards: "Almost a hundred countries have proven gas reserves adequate for commercial exploitation" (Tempest, 1986; see also Chapter 1 in this volume).

Furthermore, natural gas fits better into the increasing complexity of our industrialized infrastructures, as flexibility through integration becomes more and more important. In other words, oil technologies appear to have reached, at least in the industrialized countries, their maximum maturity within the technology's life cycle. This does not imply that oil's days are numbered in the near future. As *Figure 2.1* shows, substitution in the primary energy market is a slow process; it takes some 60 to 80 years for a new energy form to reach a market share of 50%. Therefore, oil will continue to be the dominant fuel well into the twenty-first century.

How can one introduce technological change into natural gas studies without drifting off into science fiction, but still grasping the essential technological prospects likely to emerge in the near, medium and far future? The technology substitution dynamics of the past provide an upper limit for the market penetration of natural gas, i.e., the slopes of the curves in *Figure 2.1* serve as a measure of how fast infrastructures and societies can adopt a new form of energy at a macro level. Actually, natural gas has been contributing to the global energy mix for quite some time – though, before the 1970s, primarily in North America – and thus is not an absolutely new source of energy. What is important in the context of natural gas studies concerns the degree of maturity of the gas industries and their associated technologies.

So far natural gas has been the stepsister of oil. First of all, natural gas production has been a by-product of oil production. Also, until recently, most of the nonassociated natural gas reserves have been discovered during the search for oil. One major implication of this sibling relationship is that, despite the obvious physical differences, natural gas has been part of an "oil technology" rather than developing a technology in its own right.

At the end-use side, natural gas happened to fit the existing distribution and consumption infrastructure – the Old Town gas era. Without major infrastructural adaptation, it was economically possible to market such an oil by-product. In addition, during the 1960s the Western European natural gas market received its initial growth momentum from the cheap gas of the large Dutch Groningen field. Without the Groningen gas, the European gas market might not have taken off until the North Sea fields were developed after the first oil price hike in 1973–1974, if at all. However, energy policies in many countries do not reflect the value of natural gas in their respective economies. Either end-use prices have been kept at extremely low levels – a fraction of those of directly competing fuels (essentially, oil products and electricity) – while strict controls simultaneously inhibited gas use in certain economic sectors, or prices were raised above competitive levels (gas premium) so as to reflect the perceived scarcity of oil and gas reserves. The latter attitude invited the Soviet Union and Algeria to enter the European gas market. This, in turn, opened a new array of supply security issues. The result, in one way or another, has been an artificial imbalance between natural gas demand and supply, which still dominates the thinking of many strategic planners.

Statistical evidence on methane technology productivity has been distorted by its close association with oil. An assessment of future natural gas prospects based on an implicit technology transfer from oil to methane, therefore, most likely misses the true potential of natural gas. As a first step, one must acknowledge and understand the geological and technological differences between oil and methane, so well outlined by Hefner in Chapter 1 of this volume.

Today, methane exploration, drilling, and production technologies suitable for obtaining gas from deeper and tight-formation fields are in their embryonic stage of development. Over time, these technologies will mature, and their productivity will increase, so as to make methane production from fields other than conventional oil domains a commercially viable undertaking. One should note that these technologies are primarily add-ons to existing infrastructures and technologies and do not imply any major technical breakthroughs. What is required is a perceptual breakthrough. As the automobile industry matured, once the step was taken from attaching an engine to a horse cart to an integrated automobile design, methane technology will gain momentum once it becomes emancipated from oil technologies.

Similar evolutions will also be experienced at the levels of gas transmission, distribution, and end-use conversion, which will further enhance the efficiency of gas compared with other fuels, contribute to the improvement of vital environment conditions, and function as a chemical tether for industrial development and integration. Meanwhile, two major energy-consuming sectors have traditionally been ignored in methane marketing strategies – electricity generation and the transport sector. The former market has been foreclosed through institutional regulations in many countries, while the latter suffered from an economically unviable methane-to-synthetic gasoline conversion technology.

In the long run, both markets offer opportunities for methane accommodation. Environmental considerations and sociopolitical acceptance problems associated with nuclear power in particular, and multimillion dollar projects in general, may pave the way for the entry of gas into the electricity generating sector. New technologies, such as highly efficient multistage combined cycle turbines or fuel cells in the kW to MW range, could be instrumental in a shift from centralized to decentralized electricity cogeneration systems. These combined cycle units appear to have relatively flat economics-to-scale prospects, which implies increased flexibility in unit sizes. At times of highly uncertain electricity demand projections, this is important for utility capacity planning. Small unit sizes, available virtually off the shelf, and short construction periods of six months or so, would significantly reduce front-end capital requirements and thus decrease the risk of new utility investments.

Without an efficient methane-to-gasoline conversion process, methane use in nonstationary applications will remain inferior to oil-based motor fuels. Although much research activity is targeted toward developing a direct (catalytic) conversion process, additional efforts are required before commercial viability can be achieved. Such a technical breakthrough would have a favorable impact on the prospects for natural gas in developing countries in terms of both domestic energy supplies and gas exports, simply because the transport economics for a liquid (at ambient temperature and pressure) would correspond to those for conventional oil products. A vivid example is the economic failure of the New Zealand methane-to-gasoline effort as a consequence of the indirect (two-stage) conversion process. Since both stages (gas-to-methanol and catalytic methanol-to-gasoline conversion) are associated with conversion losses (overall thermal efficiency of 54%), the economic formulas for profitability did not produce sufficiently attractive results, despite relatively low natural gas prices.

In summary, most energy projections avoid dealing with the impact of increasing knowledge on resource availability or on energy prices. Indeed, projecting the dynamics of productivity performance or innovation is a risky undertaking, but it is a necessary step if we are to understand longer-term energy demand and supply. The omission of innovation and new technologies explains, at least in part, the poor track record of conventional energy studies. The following section compares these conservative and risk-averse approaches with an attempt to introduce technical progress in a long-term gas study.

4.3. The Impacts of Technical Change on the European Gas Market

4.3.1. Technology and policy

The European gas market is characterized by a highly uncertain demand outlook, which, to a large extent, is a direct consequence of even more uncertain supply prospects. Medium to longer-term gas supplies, though plentiful in terms of available reserves and resources, are either only marginally attractive (given current energy price profiles and state-of-the-art technologies) or stem from politically diverse sources.

The agreement between Statoil and a European consortium led by Ruhrgas AG, signed in spring 1986, to develop the highly capital-intensive Troll field could have stabilized the supply outlook, had the timing not coincided with the rapid decay of world oil prices. Since Ruhrgas has not been receptive to a gas-to-gas competition (cheaper gas deliveries could have been made available by the Soviet Union), it turns out that the "commercial realities" (Stern, 1984) do not alone determine the levels of domestic gas production and dominate the negotiations of gas import contracts. Actually, Ruhrgas used the bulk availability of Soviet gas at attractive prices both as a negotiating card in the Troll contract talks and as an incentive for the

Netherlands to extend its gas export contracts for at least another 10 years. The threat of Soviet gas supplies changed Norway's expectation from that of a premium price (for the high-cost but secure supplies from West Europe) to market-related prices. Indeed, the Netherlands also extended its major export obligations for another 10 years. Thus, "noncommercial realities", i.e., national macroeconomic objectives and gas resource policies, foreign trade relations, geopolitics, and supply security concerns, overshadow the technically feasible and economically viable market potential for natural gas in Western Europe.

Therefore, apart from technical improvements, the competitiveness of natural gas in various energy markets will depend on a variety of related economic and institutional factors. Most apparent are the problems of the large initial investment in the methane transport infrastructure, the terms of actual sales contracts, gas pricing principles, take-or-pay clauses, etc; but probably as important are such factors as the political acceptability of gas deliveries from certain regions or gas purchases motivated by considerations other than volumes and prices. Generally, contracts including all these factors must be signed prior to any investment in the gas production and transmission system (as was the case with the Troll agreement). Once a contract has been signed, both producer and importer are locked into this arrangement for as long as 20 to 30 years, as a result of the enormous start-up capital requirements and the long payback periods associated with significant natural gas projects. Though viable in a purely economic calculation, private investors are reluctant to assume the economic risk involved, or the absolute level of capital outlays exceeds the financial capability of individual utilities. Consequently, government involvement has become the rule both for guarantee purposes and resource management. Hence, gas markets have become more and more regulated at a time when there has been a converse tendency – a growing spot market – for oil and coal.

Government regulations continue to affect national, as well as international, gas trade. The recent deregulation of the US gas market is a reaction to the adverse effects of the price controls imposed by the Natural Gas Policy Act. In essence, this act created a split gas market. Pipelines crossing state boundaries, as well as the wellhead pricing of such transported gas, were subject to federal regulation and jurisdiction. Hence, if a producer sold gas to an interstate pipeline system, the price was regulated. If sold within the state, the price was determined primarily by local competition. The net effect was ample supply of relatively cheap local gas and a regulation-induced shortage of cheap interstate gas. In 1978 efforts to rectify this distorted situation led to a partial regulation of the intrastate market. In 1985 gas from certain vintages and deep or tight formations was decontrolled, which was the first of further deregulation steps.

The net effect of these institutional constraints is summarized in the current conventional wisdom that European natural gas demand is more price-elastic with respect to rising than to falling prices (Purvin and Gertz, 1983 and 1985; Rogner *et al.*, 1985; Blakey 1986). Clearly, the gas delivery contracts and their underlying pricing principles reflect: (a) the objective of supplying gas at prices competitive with alternative fuels, (b) the costs of avoiding the use of alternative fuels or, in some extreme cases, (c) oil price parity at the point of production. Whatever motivation may be applicable, the determination of the sales price for gas ranges at the upper end with respect to burnertip competitiveness, given the existing technical environment. Any rise in prices, therefore, would cause a stronger response than a corresponding decline.

4.3.2. Assumptions of the IIASA study

An IIASA gas study (Rogner *et al.*, 1986) assessed the technoeconomic potential for natural gas in Europe, assuming a gradual relaxation of most internal (i.e., national or regional) policies and institutional constraints that effectively distort the competitiveness of gas in various energy markets. These distortions include, for example, banning gas from certain applications (electricity generation) or protectionism of electricity versus gas in end-use markets or prohibition of active gas-to-gas competition. Obviously, the discontinuation of these distortions creates a more favorable position for continued oil (and coal) substitution.

Current geopolitical and supply security considerations that affect import volumes from some politically diverse sources served as guidelines – in the short run – for the natural gas trade flows, but not as rigid limits or import ceilings.

Since institutional limits are difficult to project, one of the prime objectives of the IIASA study concerned the impact of a changing technical environment on the future prospects for natural gas. Consequently, technical progress constituted one of the most crucial scenario parameters in this study. Indigenous resource extraction costs and market prices were based on marginal (normalized) costs plus royalties (profits) and government take. We further assumed no link between domestic oil or oil product prices and methane prices. Import prices for crude oil, oil products, and coal were exogenously determined, while gas import prices were the combined result of a weak link to oil prices and the concept of long-term marginal resource rents.

4.3.3. The Conventional Technology Scenario

IIASA's Conventional Technology Scenario represents a business-as-usual projection in which technical progress is assumed to follow trends that have emerged since the first oil price hike in 1973–1974. In particular, the Conventional Technology Scenario includes productivity improvements at the user end of the energy system, such as industrial boilers, furnaces, etc., and residential–commercial space heating technologies and appliances. One should note that energy savings measures, such as better energy management or the existing standards for the insulation of the housing stock, have been accounted for in the useful energy demand scenario calculations. By the same token, the industrial energy demand assumptions already reflect the effects of anticipated economic–structural changes.

Because technical progress is not restricted to one particular fuel or technology, and so as not to discriminate between fuels, a uniform efficiency growth rate was applied to each category of end-use technologies. Roughly, efficiency increases for thermal conversion purposes of some 25% in the residential and commercial sectors, and of 15% in industry, were applied.

Burnertip competition was the decision criterion in the calculations for the optimal-cost primary energy mix. Given the current glut in all European energy markets, a "demand creates its own supply" situation was assumed for the short run. Competition between energy imports and indigenous production becomes a central issue where vintage structure and sunk capital cost considerations play a dominant role. Consequently, within the infrastructure of the energy system, fuel switching or interfuel substitution is primarily a function of relative prices.

In the long run, energy supply appeared not to be resource-constrained, and a least-cost supply strategy based on essentially state-of-the-art technologies was pursued. *Table 4.1* shows the international oil market price development underlying the IIASA analyses.

Table 4.1. Oil price development in the Conventional Technology Scenario, 1980–2030.

Currency	*1980*	*1985*	*1990*	*1995*	*2000*	*2010*	*2020*	*2030*
US$ 1985/bbl	39.8	27.8	22.5	25	27	33	44	49
ECU 1985/bbl[a]	45.1	31.5	25.5	28	31	38	50	56

[a]ECU = European currency unit.

The results of the IIASA Conventional Technology Scenario are summarized in *Figure 4.1*. The West European primary energy mix in *Figure 4.2* corresponds to the findings of most other energy studies, such as *Energy 2000* (Guilmont *et al.*, 1986) for the Commission of European Communities (CEC) and a study by the Economic Commission for Europe (ECE; 1986), as *Table 4.2* makes clear. Unlike the IIASA, CEC, and ECE projections, which account for the dynamics of the entire primary energy balance, a study by the International Energy Agency (IEA; 1986) reports only future natural gas developments for Western Europe. However, the evolution of the primary energy system underlying the IEA gas analyses by and large resembles the CEC and ECE projections.

When comparing the numerical findings of these studies, one should bear in mind that there are some geographical disparities in their regional aggregations. The CEC study includes the EC-10 members, the IEA study encompasses OECD-Europe, the ECE aggregation accounts for all West

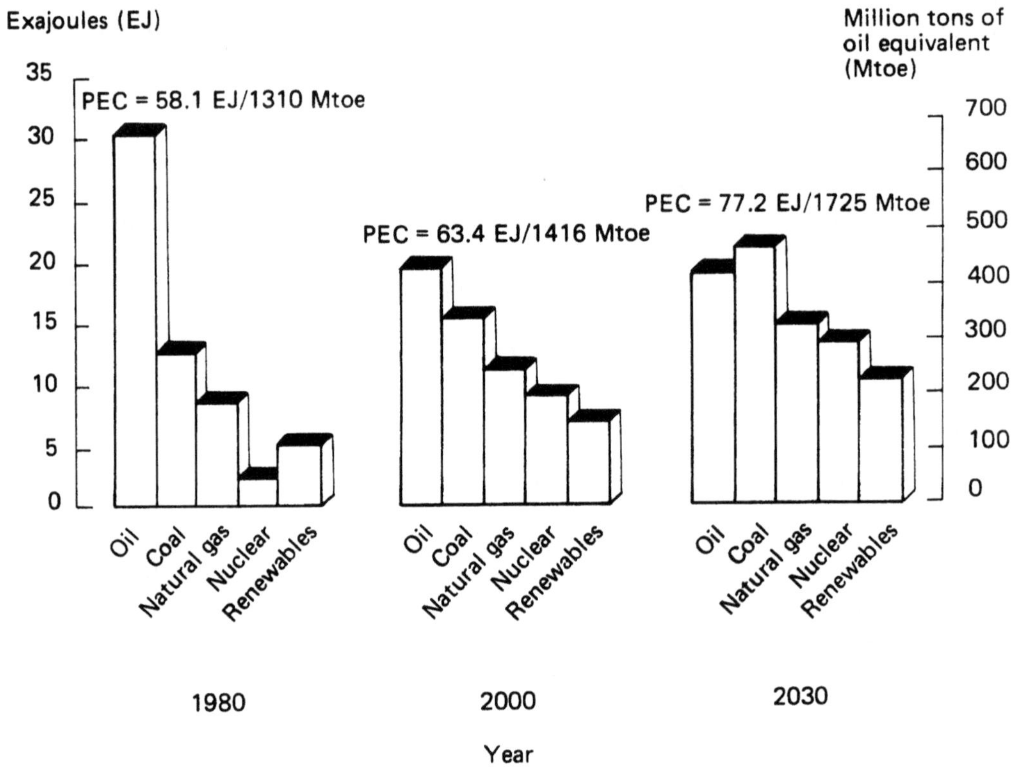

Figure 4.1. Western Europe, Conventional Technology Scenario: primary energy consumption (PEC), 1980–2030.

Table 4.2. Primary energy mix of major energy studies (in %).

	1980[a]			*2000*		
Fuel	*IIASA*	*ECE*	*CEC*	*IIASA*	*ECE*	*CEC*
Oil	52.7	53.0	53.6	35.0	35.1	38.6
Coal	21.3	21.2	23.0	23.5	23.4	23.2
Natural gas	14.3	14.7	17.5	17.3	16.9	17.3
Nuclear	3.8	3.9	4.4	13.6	19.3	18.9
Hydro and other	8.0	7.2	1.6	10.6	5.3	1.8
Total	100.0	100.0	100.0	100.0	100.0	100.0

[a]Differences in geographical aggregation and conversion factors account for the varying baseline data of 1980.

European countries, while the IIASA regional aggregation adds Yugoslavia to Western Europe. Therefore, any discussion of, and comparisons between, these studies will be restricted to relative terms, such as the evolution of the primary energy mix in terms of percentage shares, etc. Furthermore,

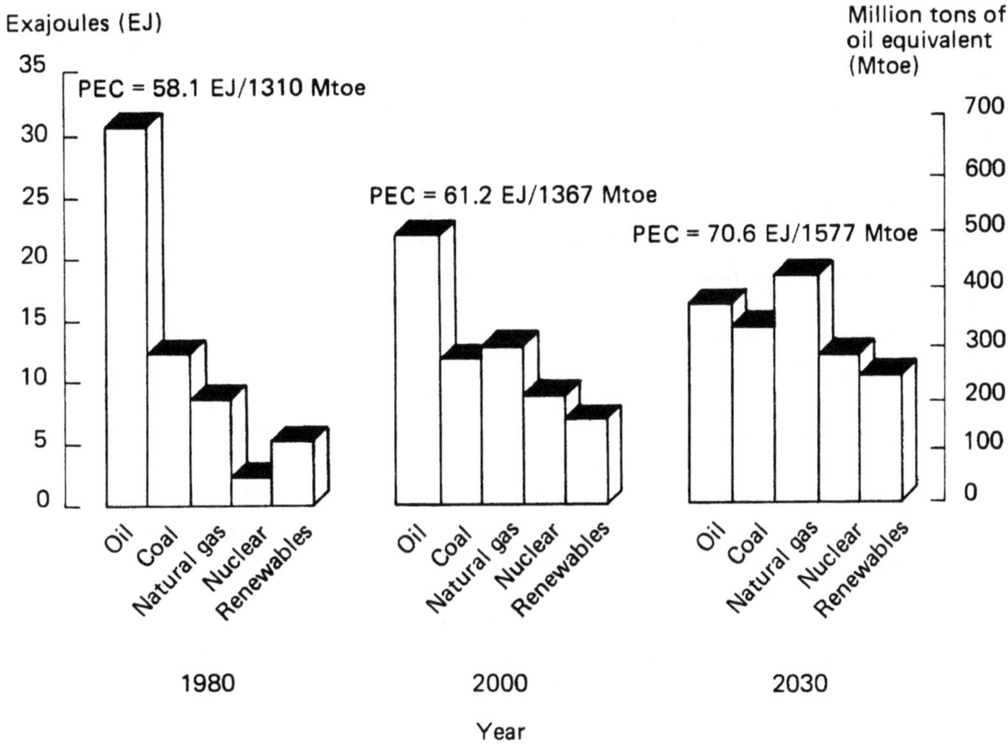

Figure 4.2. Western Europe, Technical Evolution Scenario: primary energy consumption (PEC), 1980–2030.

the IIASA study horizon extends to the year 2030, the IEA study ends in 2010, while the CEC and ECE projections do not look beyond the turn of the century. Therefore, the year 2000 will be used as a reference year for the comparison of future primary energy trends.

The common findings of these projections include:

(1) Substantial displacement of oil to less than 40% by the year 2000.
(2) Significant increase in the contribution of nuclear power.
(3) A constant share for solid fuels (hard coal and lignite).
(4) Constant to, at best, slightly increasing shares for natural gas.

Dissimilarities among the results of these studies relate primarily to the evolution of primary energy consumption in absolute terms, the rate of expansion of nuclear power, and the future contributions of renewable energy sources. The IIASA primary energy growth rates fall behind those of the CEC, ECE, and IEA as a result of different expectations regarding future economic growth rates, the assumed rate of structural change, and improvements in energy performance. For example, the CEC study bases its energy consumption projections on a gross domestic product (GDP) expansion of some 2.8% per annum for the period 1984–2000 (cf. IIASA's 2.3%). In contrast to a lower economic growth expectation, the IIASA study assumes more improvement in energy intensity than the CEC. The combined effect of these assumptions translates into diverging primary energy growth rates or primary energy–GDP elasticities of 0.22 in the IIASA study and of 0.51 in *Energy 2000*, respectively.

In the IIASA study, the contribution of nuclear power does not reach the market shares achieved in the ECE, IEA and CEC analyses. Two major reasons account for this difference, beginning with the lower level of economic activity underlying IIASA's projections. Second, the IIASA study was completed in fall 1986, and thus accounted for both recent changes in public attitudes toward nuclear energy and the generally lower energy price profiles that prevailed one or two years earlier.

Also of interest is the distribution of methane use between various energy-consuming sectors, the dynamics of which take different directions in these studies. While both the ECE and CEC projections foresee a significant decline of gas as a fuel for electricity generation by the year 2000 (see *Table 4.3*, which depicts reductions of 5.6% and 7.1%, respectively), the IIASA and IEA studies do not anticipate such a strong reduction in gas use for this sector (minus some 3.5% points). ECE, IEA, and CEC project growing gas reliance in the residential and commercial sectors, which offsets the decline in the electricity sector (and, in the cases of the IEA and CEC, in the industrial sector as well); IIASA predicts a shift from the residential–commercial to the industrial sector.

Explanations for these diverging trends can be derived from:

(1) The low GDP–primary energy elasticity of the IIASA study, which implies an even stronger restructuring of the industrial sector and a shift to more efficient and more flexible industrial boilers and furnaces than assumed in, for instance, the CEC study.
(2) The explicit inclusion of energy consumption density categories and category-specific natural gas distribution costs in the IIASA study, which may have discriminated against gas use in low-density areas.
(3) The general tendency of the IEA and CEC to favor coal over oil and gas in the industrial and central conversion sectors.
(4) No energy carrier being barred from any specific energy sector in the IIASA study.
(5) The longer time horizon of the IIASA study, until the year 2030, which automatically includes a different (larger and less geopolitically constrained) energy resource availability concept.

In summary, the IIASA Conventional Technology Scenario replicates, on the whole, the qualitative projections of the ECE, IEA, and CEC. By and large, this is not unexpected. The basic assumptions regarding energy resource availability, energy production, and conversion technologies or energy transportation infrastructures follow similar trends. However, on a

Table 4.3. Sectoral distribution of natural gas consumption, 1980–2010 (in %).

	1980				*1990*			
Sector	*IIASA*	*IEA*[a]	*ECE*[b]	*CEC*	*IIASA*	*IEA*[a]	*ECE*[b]	*CEC*
Industrial	33.7	38.2	34.5	35.6	36.5	33.6	34.1	32.3
Nonenergy	1.3	-	2.6	4.2	2.0	-	2.4	4.9
Residential–commercial	44.7	44.6	44.4	40.8	40.4	49.2	46.9	44.8
Transport	0.7	0.2	0.2	0.2	0.6	0.2	-	-
Central conversion	15.4	14.1	18.2	14.6	14.8	13.1	16.6	13.1
Losses	4.2	2.9	-	4.5	5.7	3.9	-	4.9
Total (%)	100.0	100.0	100.0	100.0	100.0	100.0	100.0	100.0
Total (bcm)[c]	219	214	201	203	256	244	261	228
	2000				*2010*			
Sector	*IIASA*	*IEA*[a]	*ECE*[b]	*CEC*	*IIASA*	*IEA*[a]	*ECE*[b]	*CEC*
Industrial	39.3	36.4	35.1	34.6	40.9	39.0	-	-
Nonenergy	3.2	-	2.4	5.7	4.6	-	-	-
Residential–commercial	38.9	47.5	49.9	47.4	37.0	48.5	-	-
Transport	0.5	0.4	-	-	0.4	0.3	-	-
Central conversion	12.8	11.4	12.6	7.1	11.4	8.2	-	-
Losses	5.3	4.3	-	5.2	5.7	4.0	-	-
Total (%)	100.0	100.0	100.0	100.0	100.0	100.0	-	-
Total (bcm)[c]	290	280	277	235	328	305	-	-

[a]IEA – High Demand/Low Oil Price Scenario; nonenergy use included in industries.
[b]Losses included in central conversion.
[c]Billion cubic meters.

more disaggregated level, the agreement among these studies begins to disappear. IIASA's explicit introduction of technical progress in energy conversion facilities of the industrial and residential–commercial sectors, as well as the assumption of natural gas use in all sectors unconstrained by policy changes, resulted in diverging energy consumption profiles. When technical progress is introduced as a variable in the energy supply system, therefore, we can expect further divergence of projected trends, which may eventually be felt at the aggregate level of primary energy consumption.

4.3.4. The Technical Evolution Scenario

The Technical Evolution Scenario assumes that technical progress will extend beyond end-use technologies to include all stages of the energy chain, from resource extraction to end-use conversion. Apart from improving existing or conventional technologies, this scenario also implies the availability and introduction of new exploration and production technologies, advanced equipment for the construction of gas mains, new maintenance and monitoring systems for pipelines, as well as direct methane-to-gasoline conversion, new electricity cogeneration, and other advances in energy end-use technologies. *Table 4.4* summarizes those technologies that may advance future productivity along the entire gas chain.

The most crucial assumptions in this Technical Evolution Scenario concern productivity growth rates in the exploration and drilling sectors, the methane transport system, and the availability of highly efficient methane-to-electricity conversion facilities. The assumed 4.5% annual productivity improvement for drilling in tight formations corresponds to an order of magnitude over the next 50 years that translates into a proven methane reserve growth rate of some 2% per year. (For oil, corresponding considerations result in less than 0.5% annual reserve expansion.)

Transmission and distribution costs for natural gas dominate, even in such ultra-high-cost producing areas as Troll or Sleipner in the North Sea (see *Table 4.5*). Hence, high productivity increases in the gas exploration and drilling sectors will be dissipated unless similar evolution occurs in methane transport and distribution systems. This will become increasingly important as the distance between gas resource locations and consumption centers grows. Rough estimates suggest a tripling of methane travel distances between now and the turn of the century (see also Valais, 1983). In the longer run, one might expect a tradeoff between high-cost production close to the point of consumption versus inexpensive production and long-distance transmission.

The productivity growth rate for the transmission and distribution of natural gas (see the potential improvements listed in *Table 4.4*) in the Technical Evolution Scenario amount to some 1.2% per year. Other essential productivity increases have been projected for gas-fired combined cycle turbines – see, for example, Chapter 9 in this volume. Gas turbines, a technological spin-off from aircraft jet engines, continue to benefit from the research and development efforts directed toward better turbine materials and designs. The serial combination of combustion and steam turbines currently yields efficiencies of 47 to 48% (electricity only) and more than 80% of fuel utilization in cogeneration applications. Ongoing research to reduce material degradation at high temperatures, as well as other related R & D activities, promise additional productivity gains for gas-fired combined cycles. Efficiency of more than 60% (electricity) appears to be technically feasible within a 50-year planning horizon.

The overall impact of the introduction of technical progress on the dynamics of natural gas in the West European energy market is portrayed in *Figure 4.2*. The corresponding market shares are given in *Table 4.6*.

Compared to the Conventional Technology Scenario, *Figure 4.2* reveals not only a substantially different evolution for the West European primary energy mix, but also a 10% reduction in total primary energy consumption by the year 2030 (4% by 2000). First of all, technical progress will better the overall energy system's efficiency. Secondly, one may expect a major shift in the composition of primary energy consumption, essentially from coal to

Table 4.4. **Current and future technical advances along the gas chain.**

		Productivity increases assumed[b]	
Stage	*Technologies*[a]	*Area*	*%/year*
Exploration	Satellite search, aeromagnetic mapping, carbon and helium isotope ratios, computerized geochemical and geophysical prospecting, vegetative anomalies	proven gas reserves	2.5
Drilling and production	Measurement while drilling, smart downhole electronics, polycrystallic diamond bits, improved drill pipe and casing materials, improved downhole motors, tunnel concepts, seabed platforms, gas treatment	onshore conventional onshore tight formation offshore deep water	2.0 4.5 2.5
Transport: Pipe	Large diameter pipes, higher pressures, automated pipe-laying techniques	average improvement	1.0
LNG	Magnetocaloric refrigeration	average improvement	0.75
Distribution	Ground-piercing impacters, pipe pushers, compacting augers, new accurate horizontal drilling techniques, improved maintenance and monitoring systems, pavement cutting, greater mechanization and automation, new materials	average improvement	1.25
Electricity cogeneration	Combined cycle gas turbines, fuel cells	as of 1990	0.75
Industries	Steam-injected turbines for cogeneration, heat pumps, gas desiccant cooling systems, new industrial furnaces, forced drought immersion, oxygen and natural gas burners, recuperative burners, control systems, modular concepts	average conversion efficiency	0.5
Commercial	Condensing water heaters, heat pumps, cooling systems, heat recuperation	average conversion efficiency	0.75
Residential	Heat pumps, new furnaces and water heaters, condensing and exhaust heat recovery, control systems	average conversion efficiency	0.75
Transport	CNG, catalytic methane-to-gasoline conversion	costs for methane-based motor fuels	2.0

[a]Incomplete list.
[b]The IIASA Technical Evolution Scenario uses simplifications and aggregations of new technical developments. Thus, the estimated productivity increases must be viewed as gross averages.

Table 4.5. Cost structure for methane deliveries in Central Europe (adapted from Strubegger and Messner, 1986).

	Cost Structure (%)		
Source	*Extraction*	*Transport*	*Distribution*[a]
Groningen	6	4	90
Netherlands offshore	9	6	85
Ekofisk	16	11	73
Statfjord	16	19	65
Sleipner	27	11	62
Heimdal	32	11	57
Troll	27	17	56
Gullfaks	34	15	51
Troms[b]	26	26	48
Troms[c]	25	28	47
Algeria	4	28	68
Algeria (LNG)	4	36	60
Nigeria	5	36	59
Nigeria (LNG)	4	43	53
Persian Gulf (pipeline via Yugoslavia)	4	32	64
Persian Gulf (pipeline via Italy)	3	35	62
Persian Gulf (LNG)	3	44	53
Urengoy	8	34	58
Other Western Siberia	10	35	55

[a]Distribution costs have been averaged and kept constant for all cases.
[b]Onshore pipeline via Sweden to Hamburg.
[c]Offshore pipeline to Emden.

Table 4.6. Primary energy mix, Technical Evolution Scenario (TEC) and Conventional Technology Scenario (CONV), 1980–2030 (in %).

		2000		2030	
Fuel	*1980*	*TEC*	*CONV*	*TEC*	*CONV*
Oil	52.7	35.6	35.0	22.6	24.3
Coal	21.3	19.2	23.5	20.3	26.8
Natural gas	14.3	20.5	17.3	25.9	18.9
Nuclear	3.8	13.8	13.6	17.0	16.7
Hydro and other	8.0	11.0	10.6	14.2	13.2
Total	100.0	100.0	100.0	100.0	100.0

natural gas, which will further improve energy system performance. Finally, natural gas will become the largest single primary fuel in Western Europe by the year 2030 (market share: 26%).

The evolution toward a gas-dominated European energy system begins to emerge by the turn of the century, though not repeating the growth dynamics historically observed with oil:

(1) The market share of natural gas will exceed the 20% mark.
(2) Coal will lose almost 4% of its market share and decline to a historical low (since the industrial revolution) of less than 20%.
(3) Oil's displacement will be affected only marginally by the assumptions incorporated in the Technical Evolution Scenario.

The gross consumption of natural gas expands from 220 billion cubic meters (bcm) in 1980 to 330 bcm by 2000, which is an increase of 43 bcm versus the Conventional Technology Scenario. Because of the productivity gains underlying the Technical Evolution Scenario, these additional natural gas volumes will not translate into corresponding increases in gas imports. In fact, the overall gas import dependence from non-European sources will be reduced by some 30 bcm to 95 bcm by the year 2000. Similarly, oil imports are expected to decline by 6%; coal imports, by more than 50%. Technical change will not only reduce the absolute level of oil import requirements, but also exert pressure on the international oil market price. The oil price development shown in *Figure 4.3* reflects the combined effects of the increased gas market share, the reduced demand for oil imports and the positive impact of technical change on the cost profile of European oil production. [The price of natural gas imports from non-European sources can be expected to fluctuate between 70% and 80% of the international oil market price (fob at the European entry point).]

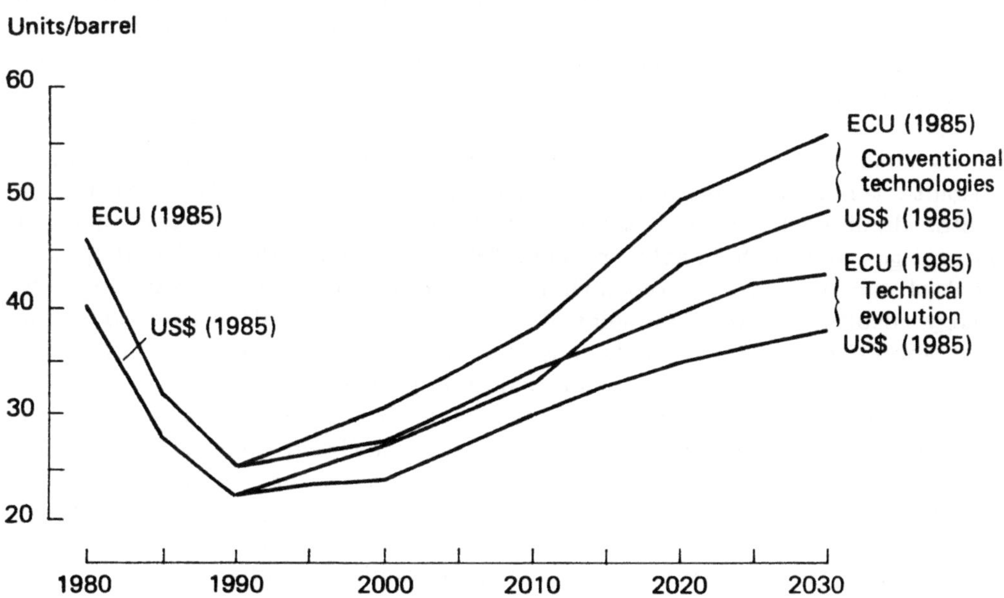

Figure 4.3. Oil price development in the Technical Evolution Scenario versus the Conventional Technology Scenario, 1980–2030.

After the year 2000, the market penetration of natural gas continues at the expense of coal and, to a lesser extent, of oil. In 2030 aggregate gas consumption amounts to some 480 bcm, 124 bcm of which would originate from non-European sources. This implies an indigenous West European gas production of some 350 bcm or a doubling of current production levels. (In the Conventional Technology Scenario, gas imports amount to 135 bcm and indigenous production to 153 bcm.) The projected production level of 360 bcm is the result of combining assumptions about technical progress for methane exploration (i.e., the resource base) and drilling (i.e., tapping the resource base). These somewhat unconventional views on technological evolution translate into economic conditions that will favor gas over other fuels. In other words, the methane production projections of the Technical Evolution Scenario require the existence and the technoeconomic feasibility of tapping gas resources, such as Devonian shales, tight sands, pure pressure aquifers, gas hydrates, subducted deep deposits, and other abiogenic sources. For example, Hanneman (see Chapter 7 in this volume) estimates US unconventional gas resources at over 250 trillion cubic meters (tcm), versus the current official gas reserve figure of some 5.4 tcm. Again, one should note, no one suggests that these vast resources could be recovered with today's technology.

The sectoral uses of natural gas in both scenarios are depicted by *Table 4.7*, in which total natural gas consumption does not include losses in the transmission and distribution system (approximately 5%). A comparison between the Conventional Technology and Technical Evolution Scenarios points to two reciprocal consequences of technical progress. These are the efficiency improvements introduced by new advanced technologies and the absolute level of fuel consumption. It appears that the efficiency improvements outbalance the additions in actual market shares. For example, in absolute terms, consumption of natural gas in the residential–commercial sector declines by some 12% from 108 bcm to 95 bcm, while the share of natural gas within this sector remains fairly constant. Hence, the displacement of one fuel unit of conventional technology by a more advanced technology results in lower fuel sales – a fact many gas utilities learned in the past when their sales forecasts, based on a one-to-one oil substitution strategy, failed to result in corresponding actual gas sales.

The industrial sector is less sensitive to this phenomenon than the residential–commercial sector, and absolute methane consumption should decline only marginally. New technologies will allow for additional gas consumption in the low-temperature market, while electricity will replace gas in certain areas of high-temperature applications. The latter is a comprehensive process, rather than an energy system-specific peculiarity.

Considerable growth in natural gas consumption is expected to occur in the central electricity and heat (cogeneration) sector and the transport sector. Both sectors are projected to experience more than a twofold increase in their gas consumption, although actual gas volumes will differ greatly. These developments indicate a distinct structural change in the sectoral pattern of natural gas consumption. Natural gas as a final energy

Table 4.7. Sectoral distribution of natural gas consumption.

			2000				2030			
	1980		TEC[a]		CONV[b]		TEC		CONV	
Sector	%	bcm	%	bcm	%	bcm	%	bcm	%	bcm
Industries	35.2	72.6	34.0	106.3	40.2	108.5	28.5	124.0	43.7	160.4
Feedstocks	1.4	2.9	3.0	9.3	3.4	9.2	8.7	37.9	9.0	33.0
Residential–commercial	45.4	93.6	30.4	95.1	39.9	107.7	23.7	103.2	37.0	135.7
Transport	0.7	1.4	1.0	3.2	0.5	1.3	6.6	28.5	0.4	1.5
Central conversion	17.3	35.7	31.6	98.8	16.0	43.2	32.5	141.2	10.0	36.6
Total	100.0	206.2	100.0	312.8	100.0	269.8	100.0	434.9	100.0	367.1

[a]TEC = Technical Evolution Scenario.
[b]CONV = Conventional Technology Scenario.

form, i.e., as a fuel at the energy end-use level, will lose in favor of gas in embodied form:

> Already in the year 2000, and certainly more so by the end of the study horizon, any expansion of the gas market share at the level of primary energy translates into a fast penetration of natural gas into the electricity and eventually the transport sector. This also means that the industrial and residential/commercial markets offer little room for additional gas penetration. In both markets (burnertip) competition from district heat and electricity limit the use of methane. Instead, natural gas finds a growing niche as a valuable fuel for electricity and heat generation.

The percentage distribution of gas use shown in *Table 4.7* confirms this projected trend. In the short run, the relative expansion of gas use for electricity generation occurs at the expense of the industrial and residential–commercial shares. Both sectors continue to lose market shares throughout the study horizon; but the transport sector, instead of the electricity sector, will gain the bulk share. Here, the assumption of a technically feasible and commercially viable methane-to-synthetic gasoline conversion process is a necessary prerequisite.

4.4. Conclusion

The Technical Evolution Scenario shows that incorporating technical progress into long-term energy scenarios has a considerable impact on the future prospects for the West European energy system. In other words, major opportunities for an expanded use of natural gas hinge on the successful development of new gas utilization and supply technologies.

Although forecasting technological change is as risky as any other predictive exercise, there are definite benefits from including technical change in medium- to long-term energy analyses. The principal benefit of such an exercise is to challenge conventional wisdom, which tends to extrapolate the status quo of technical and economic conditions into the future. Although extrapolation is the essence of the Technical Evolution Scenario, an attempt was made to assess the dynamics of technologies and study their consequences for the future market penetration of natural gas.

Another important factor concerns the mutual influence of energy policy and conventional wisdom. For example, the myth of the 1970s averred that natural gas, as a scarce resource, would be worth more kept in the ground than released into the current marketplace. This perception changed with the discovery of major gas fields of the North Sea, particularly the Troll field, and the introduction of Soviet and Algerian gas into the European gas market. By the same token, the benign environmental characteristics of natural gas have been used to negotiate parity prices, at least, on equivalent calorific terms with oil. Technology, therefore, could be the vehicle to overcome these myths, turn submarginal resources into economically recoverable reserves, and thus pave the way for natural gas to become a globally dominating fuel.

Given the current energy bubble, the momentum for technical change must occur at the level of gas-utilizing technologies. This is to say that, in the short run, improvements of existing boilers, furnaces, and gas turbines, as well as the simultaneous development of new gas-consuming devices, cogeneration facilities, or gas-based industrial processes, will be the stimulus not only for increased gas consumption, but eventually also for the anticipated technological improvements on the gas supply side.

The IIASA gas study estimates a market share of 26% for gas by the year 2030. Compared with the technology substitution model, depicted in *Figure 2.1*, this is less than 50% of the market share derived from historically observed substitution dynamics. As a matter of fact, in the past the market penetration of methane has been slower than, for instance, the displacement of coal by oil. Again, this observation points to the historical situation of gas as a dependent fuel, manufactured initially from coal and later as a by-product of oil. Current gas technologies represent either old or oil-based technologies.

To that extent, the technical progress assumptions underlying the Technical Evolution Scenario may be too conservative, in particular, for those markets where the expansion of methane use encounters technical difficulties – the transport sector and feedstocks. Both sectors must eventually accommodate methane if the market share of natural gas is to exceed the 30% threshold.

The Technical Evolution Scenario is based primarily on technoeconomic considerations and omits all the equally important policy, institutional, supply security, and geopolitical factors. Combined, these factors may, or may not, impede the prospects for natural gas in the short-to-medium term. Technology may eventually mitigate these impediments. On the other hand, gas offers a viable alternative to imminent environmental problems. On all counts, methane is superior to other fossil fuels.

The purpose of this chapter was to emphasize the possible impact of future technical evolution on future energy mixes. The analysis attempts to show that the least likely event – no technical progress, which is the usual concept in long-term energy analyses – precludes intriguing alternatives *a priori*. This is not to say that the potential developments outlined here are imminent or preordained. However, the crucial question is, "How well are we protected against future technical surprise while planning continues in a constant-technology environment?" This question is of importance, first of all for energy policy and research and development at the level of national governments and international organizations. It is equally important for industries involved in long-term energy projects, such as utilities; exploration, drilling, and mining companies; and energy equipment manufacturers.

Acknowledgment

This chapter is a revised version of an article entitled "Technology and the Prospects for Natural Gas", scheduled for publication in the February 1988 issue of *Energy Policy*; it appears here with the kind permission of the journal's publisher, Butterworth Scientific Limited (Guilford, UK).

References

Blakey, S. (1986), Europe's natural gas industries, in Conant, M.A. (ed.), *The World Gas Trade: A Resource for the Future* (Westview Press, Boulder, CO).

Data Resources, Inc. (1984), *The DRI Report on Natural Gas Markets in Europe* (DRI, Paris).

Economic Commission for Europe (ECE) (1986), *Energy Data Bank* (ECE, Geneva).

Gas Research Institute (1985), *1985 GRI Baseline Projection of U.S. Energy Supply and Demand to 2010* (GRI, Washington, DC).

Guilmont, J-F., McGlue, D., Valette, P., and Waeterloos, C. (1986), *Energy 2000* (Cambridge University Press, on behalf of the Commission of European Communities, London).

International Energy Agency (IEA) (1986), *Natural Gas Prospects* (Organization for Economic Cooperation and Development, Paris).

Mossavar-Rahmani, B. (1983), *OPEC and the World Oil Outlook*, *The Economist* Intelligence Unit, Special Report No. 140 (EIU, London).

Grübler, A. and Nakicenovic N. (1986), *The Dynamic Evolution of Methane Technologies*, Working Paper WP-87-2 (International Institute for Applied Systems Analysis, Laxenburg, Austria).

Peebles, M.W.H. (1985), *Prospects for Natural Gas Through to the Year 2000*, 16th World Gas Conference, IGU/TF1-85 (International Gas Union, Paris).

Prospects for Natural Gas Trade in North America and Western Europe: Reports of Two Executive Sessions (1985), Discussion Paper Series E-85-

Purvin and Gertz, Inc. (1983 and 1985), ***Western Europe Natural Gas Industry Market and Economic Analysis to 2000*** (Purvin and Gertz, Inc., Consultant Engineers, London).

Ravetz, J.R. (1984), Usable knowledge, usable ignorance: Incomplete science with policy implications, paper presented at the ***Sustainable Development of the Biosphere Task Force Meeting***, August 27–31 (International Institute for Applied Systems Analysis, Laxenburg, Austria).

Rogner, H-H., Messner, S., Strubegger, M., and Golovine, A. (1985), ***The IIASA International Gas Study***, Discussion Paper for gas meeting at Krainer Hütte, October 17–18 (International Institute for Applied Systems Analysis, Laxenburg, Austria).

Rogner, H-H., Messner, S., Strubegger, M., and Schmidt, E. (1986), ***The Methane Age***, Working Paper WP-86-68 (International Institute for Applied Systems Analysis, Laxenburg, Austria).

Stern, J.P. (1984), ***International Gas Trade in Europe: The Policies of Exporting and Importing Countries*** (Heinemann Educational Books, London).

Strubegger, M. and Messner, S. (1986), ***The Influence of Technological Changes on the Cost of Gas Supply***, Working Paper WP-86-38 (International Institute for Applied Systems Analysis, Laxenburg, Austria).

Tempest, P. (1986), Beyond the rainbow: Gas in the twenty-first century, in Conant, M.A. (ed.), ***The World Gas Trade: A Resource for the Future*** (Westview Press, Boulder, CO).

Valais, M. (1983), World natural gas prospects, ***Energy Exploration and Exploitation*** **1**(4): 269–276.

CHAPTER 5

Origin of Petroleum: Two Opposing Theories and A Test in Sweden

T. Gold

5.1. The Biological Origin Theory

Most people believe that there has been ample proof that all petroleum and natural gas (principally methane, CH_4) derives from biological substances, plant or animal debris, that has become buried and then subjected to heat and pressure. When this viewpoint was first voiced, about 100 years ago, it seemed the only possible explanation for these strange fluids that appeared so mysteriously and yet so conveniently on the surface or at shallow depths. In the context of the chemical knowledge of the day, it seemed almost inconceivable that hydrocarbons, such as petroleum and methane, could be created without the assistance of biology; they were compounds of "organic" chemistry, a term that stems from those times and denotes essentially all chemistry concerned with carbon in unoxidized form.

Also, a hundred years ago, it was thought that the Earth had once been a molten ball and then had slowly formed a crust from that melt. In that case there would be no expectation of any hydrocarbons surviving, and little chance of them being subsequently created from the melt or its frozen products. Starting from those positions, an explanation for the origin of petroleum in terms of biological debris in the sediments seemed almost compelling.

But there were dissenters, the most prominent of them being Mendeleev (1877), the great Russian chemist, whose thesis on the origin of petroleum concluded that it had come up from the depths of the Earth. He gave many arguments for this viewpoint, most of which are still valid today. Doubts about the biological origin continued to be expressed by knowledgeable people, especially in the Soviet Union, where groups of scientists looked for and collected much evidence favoring a nonbiological origin; but the bio-origin theories seemed unbeatable, as chemical evidence piled up.

It was found that petroleum generally contained complex molecules, which were quite specific "biomarkers" ("chemical fossils"). Furthermore, some fractions of petroleum showed "optical activity", i.e., they caused the plane of polarized light to be rotated. This means that complex three-dimensional molecules, which could have a right- or a left-handed character, occur in unequal proportions. Biology introduces such asymmetries, because it creates molecules from templates that have such asymmetries built into them. Nonbiological chemistry is thought to be strictly symmetrical in this respect.

With these two observations, it seemed to many that the case was settled. The biological origin of petroleum and methane, which is so commonly associated with oil and which is just another hydrocarbon, seemed proven.

5.2. The Abiological Origin Theory

Nevertheless, problems remained. The geographic distribution of petroleum seemed to point to much larger-scale features than those of individual sedimentary deposits or, indeed, of any surface features. The quantities of oil and gas that were found turned out to be hundreds of times larger than the estimates initially made on the assumption of a bio-origin.

The viewpoint that hydrocarbons were a bio-product, which could not be expected to arise without biology, became quite untenable when it was clear that hydrocarbons are in fact the most common form of carbon in the entire planetary system. Jupiter, Saturn, Uranus, and Neptune contain enormous amounts of methane and other hydrocarbons in their giant atmospheres. Titan, a satellite of Saturn, has an atmosphere consisting of dense clouds of methane and ethane, much like the Earth's clouds of water vapor. The comets and the asteroids contain various types of hydrocarbon compounds. The core of comet Halley, recently investigated, was found to have a surface that was "black as pitch" – most probably because it was made of pitch or a similar hydrocarbon substance. All this material is most unlikely to have been created by biological processes.

The idea that the Earth had formed as a ball of liquid rock has long since been replaced. We are now certain that the Earth accumulated from solids, and that only partial melting caused the formation of a relatively thin crust of lighter rocks, segregated from the largely solid mantle that forms the outer half of the radius, or seven-eighths of the planet's volume. In this picture it can no longer be ruled out that volatile fluids could have been cooked out from this mantle and worked their way up toward the surface.

The water of the oceans and the nitrogen of the atmosphere, as well as smaller quantities of other volatile substances of the outer crust, must have originated from such an "outgassing" process. The element carbon is abundant on the surface of the Earth and in the sediments in oxidized form, as carbonate rocks, and unoxidized, as coal and other carbon compounds; but carbon is only a quite minor component of the basement rocks that generated, through erosion, the bulk of the sediments. Carbon must have entered the crust from deeper levels as a fluid to provide that great quantity of "carbon excess". Moreover, it seems to have come up over most of the timespan of the geologic record, since the proportion of carbon in sediments has greatly increased from the oldest sediments to the youngest. Carbon dioxide and methane are the principal candidates for the fluids that could have brought up all this carbon. In the highly oxidizing atmosphere, carbon would in any case end up chiefly as carbon dioxide.

The meteorites represent a small amount of material left over from the time of formation of most of the planetary bodies. They provide a great deal of evidence for the substances and their chemical condition in the formation process. Among all the types of meteorites, only one class – the carbonaceous chondrites – could be held responsible for bringing much carbon to the Earth. This material, when subjected to heat and pressure, after being deeply buried in the forming Earth, would produce fluid hydrocarbons, probably chiefly methane, and also smaller quantities of CO_2. It is likely, then, that the bulk of the carbon that the Earth acquired, and from which it derived its surface carbon excess, was initially supplied in the form of heavy, almost solid hydrocarbon compounds, such as those in carbonaceous chondrites. The old viewpoint that hydrocarbons could be on the Earth only as a result of biology no longer has any basis.

Diamonds have been brought up by eruptions from great depths. They represent a pure form of carbon, and the pressure necessary for their formation occurs only at a depth of 150 km or more. It is clear, therefore, that unoxidized carbon in high concentrations does exist down there. Other forms of unoxidized carbon can also be expected to be present there, and indeed methane has been detected in pore spaces within diamonds.

For a long time, it was thought that the high temperatures at these depths would dissociate methane and other hydrocarbon molecules. This has been contradicted by observations and also by modern thermodynamic calculations, which recognize the important stabilizing effect exerted by the enormous pressure. In fact, it is now suggested that a mix of the principal types of molecules that make up petroleum would be expected in such a pressure–temperature regime, if carbon and hydrogen were available (Chekaliuk, 1976). It would be useful to be able to demonstrate this in laboratory experiments, but the combination of such extreme values of temperature and pressure has so far not been achieved.

We suppose now that the cooking out of carbonaceous chondrite material provides carbon and hydrogen, and therefore that methane loaded with heavier petroleum molecules results and begins its ascent from these levels. If the ascent takes it through pathways where liquid magma exists, then the oxygen available in the magma will largely oxidize these fluids, and carbon dioxide and water will rise to the surface. If, however, the pathways go through cracks in solid rocks – cracks that are forced open by the high pressure of the fluids – then the oxygen available on the crack surfaces will soon be exhausted, and methane and petroleum can reach the shallower levels without being destroyed. At shallow levels, pore spaces will tend to be filled by such upwelling streams; and because much more porosity exists in sediments than in any other rocks, one will find oil and gas concentrated there.

The "molecular fossils" and other biological molecular evidence in petroleum do not indicate more than a biological contamination; they provide no evidence for the origin of the bulk substances. Contamination arises from (1) bacteria that feed on the oil, and whose cell wall molecular compounds (hopanoids) have, in fact, been identified in most oils; and (2) from the contaminants that oils pick up as they migrate through organic debris, which is frequently present in the sediments.

Had these facts been known earlier, the bio-origin theories would not have been adopted as firmly, and one would have hesitated to build enormous structures of detailed explanations upon them. Some chemists continued to voice doubts; they noted that petroleum was largely hydrogen-saturated, whereas they would have expected a hydrogen deficiency in buried biological materials. The evidence of petroleum seemed to them to point more toward biological contamination of an abiogenic substance, rather than to a biological source material. Sir Robert Robinson (1963 and 1966), the distinguished chemist, Nobel Laureate, and President of the Royal Society, wrote:

> Actually it cannot be too strongly emphasized that petroleum does not present the composition picture expected from modified biogenic products, and all the arguments from the constituents of ancient oils fit equally well, or better, with the conception of a primordial hydrocarbon mixture to which bioproducts have been added.

But perhaps the greatest problem for the bio-theories was the association with helium. Many of the petroleum and methane-bearing areas of the Earth are also outstanding for their helium richness. All commercial helium is derived from natural gas. Very few areas on the Earth have high concentrations of helium in their underground gases without methane being the dominant partner. This relationship is inexplicable if the methane had been derived from bio-materials buried in the sediments. Biology can have had no part in the process of concentrating helium, a chemically quite inert gas. The explanation that helium just happens to be caught in the same traps that hold methane is inadequate, when one realizes that there are very many traps that do not happen to have methane in them. Why do these traps lack high proportions of helium?

Global patterns of hydrocarbon occurrence also continued to challenge the bio-theories. Why was the Middle East so rich in hydrocarbons? The areas of the mountains of southeastern Turkey, the valley of the Tigris, the folded mountains of Persia, the Persian Gulf, and the flat plains of Saudi Arabia have little in common with each other, except that they form one connected region that is enormously well supplied with oil and gas. No unifying feature for the region as a whole has been discovered. The oilfields not only span different geologic ages, but they have different reservoir rock and different caprocks holding down the contents of those reservoirs. The attempts to find sediments rich in biological debris have generally failed, and there is no consensus as to the source material that produced all this wealth of oil and gas. Kent and Warman (1972), commenting on their detailed geologic studies of the area, wrote:

> It is a remarkable circumstance that the world's richest oil-bearing region is deficient in conventional source rocks.... The oil is distributed in reservoirs varying in age from Middle Jurassic to Miocene.... Despite this range of age and type of reservoir there is a notable homogeneity in the chemical composition of the oils, and there is a presumption that they have a common stratigraphic origin.

It has also been noted for the entire region that "Most of the reservoirs are conspicuously full to the spill level". Where could this huge quantity of source material be hiding in a region that is by now very well explored? Can one really suppose that a single source region could have delivered its product by horizontal transport in the sediments to all the diverse deep and shallow fields of the entire area? Or was it accidental that pockets of different source materials, laid down in quite different circumstances, produced oils of notable chemical homogeneity?

In many other parts of the world, one can also recognize geographic patterns where much oil and gas is found, spanning different geologic settings. The vertical stacking of hydrocarbon deposits has also been noted: areas that have oil or gas at shallow levels often have more at deeper levels, sometimes all the way down to, or even into, the basement rock. If deep-source materials were responsible for the deposits, one would judge that these are distributed unevenly, in large-scale patterns that have no simple relationship to the crustal features. How much oil and gas accumulates in an area would then depend on the magnitude of this supply from deep levels, and also on the faults and other tectonic features that may facilitate the upward migration. In an area such as the Middle Eastern oil provinces, the supply seems to be prolific, and every trap has been filled, irrespective of the local geologic setting. The vertical stacking of deposits must clearly be expected in such a picture.

Many other features of the occurrence of oil and gas can then be understood. Helium, mostly produced by radioactivity in the rocks and normally present in very small concentrations only, would be picked up by gas that travels a long way through cracks in the rocks. The striking association of helium with methane then suggests that methane has usually traveled over great distances, presumably because it has come up from great depth. Other associated trace constituents that are volatile, such as mercury, can be similarly accounted for. Chemical relationships between various forms of unoxidized carbon that are found in the same region, some of them thought to be of plant origin and given the name of kerogen, can be understood if they all derive mainly by precipitation from a common upwelling stream of methane and other hydrocarbons from deep sources.

The concentration of petroleum filling all available pore spaces in some porous rocks was always thought to require a transporting agent. If that transporting agent were methane, which at high pressure is a good solvent that will thin and mobilize oil, then the quantities of methane required for this would have been orders of magnitude larger than those now in the ground, or estimated to be derivable from any organic sediments. If deep-source methane has in fact been responsible for bringing up most of the carbon excess of the surface of the Earth, then such a large supply must have existed in any case. Petroleum transported in the methane would tend to be shed out of solution as the pressure drops in the upwelling gas, and as the properties of methane are thereby changed from those of a high-density liquid to those of a low-density gas.

5.3. Testing the Theories in Sweden

Despite the apparent strength of all these arguments, it was clear that the firmly established viewpoint of the biological origin of all of the Earth's hydrocarbons could not easily be displaced. An example of a large hydrocarbon occurrence in an area where there was little or no chance that biological deposits could have contributed would clearly have the greatest impact. The ancient granitic shield of Scandinavia seemed a prime candidate. If hydrocarbons seep up from the mantle below, then the granite of Sweden would appear to be resting on a particularly hydrocarbon-rich area of the mantle. The Norwegian Trench on one side, with its numerous oil and gas fields, and many hydrocarbon occurrences, as well as some commercial production to the east and south, could be understood in that way. It was also known that many cracks in the granite of Sweden have deposits of tar in them, which may also be due to hydrocarbon seepage from below. If any large region of porosity existed in the granitic rock of Sweden, it might be expected to contain gas and oil.

It was on this basis that the author made contact with the Swedish State Power Board, where an influential group (under the direction of Dr. Tord Lindbo) was willing to pursue these ideas. Among the areas where porous rock might be found, the outstanding candidate was Europe's largest meteorite impact crater, situated in Central Sweden and known as the Siljan Ring. When investigations were started there, it was quickly discovered that much evidence already existed, pointing to just this structure as an area of seepage of both gas and oil. Farmers' water wells had long been known in the area to produce combustible gases, and numerous oil seeps were noted in the small sedimentary deposits of the ring-shaped depression that marks the crater.

After three years of investigation, it became quite clear that the area was indeed outstanding for all the effects that are associated with the seepage of gas in other oil and gas areas of the world. Anomalously high levels of methane were found in the numerous lakes and ponds of the area. Carbonates of a kind thought to be derived from the local oxidation of upwelling methane were found in characteristic patterns in the surface oil, together with trace metals that also usually accompany such seepage. Methane-derived carbonates were also found in the seven probing holes that have been drilled to a depth of approximately 500 m. Hydrogen, methane, ethane, ethylene, and propane were found in all these holes, usually in small concentrations only, but occasionally in concentrations approaching combustible levels. A gravity survey over the area showed a marked low-gravity anomaly, similar to such anomalies known over other large impact craters, and interpreted as being due to substantial porosity in the smashed rock. In the area of the maximum negative gravity anomaly, seismic profiles showed several strongly reflecting horizons, stacked one above the other, implying changes in density or sound speed at these levels. Porosity, of course, has a major effect on sound speed, and one may therefore interpret these reflections as being due to mineralization processes that have cemented up the pore spaces at particular levels, leaving different fluid pressures and different porosity proportions above and below each such layer. This information, as well as other corroborating items, suggest that fluids are trapped in this region, and that hydrocarbons are among them.

The total volume of pore space, as indicated by the gravity measurements, is very large. If filled with gas, the area would be a gas field of world class. Of course, the pore spaces may be just filled with water; but then one could not understand why the near surface showed such strong indications of gas and oil. Even a small fraction of the pore spaces filled with hydrocarbons could make it a commercially very significant field.

The Swedish State Power Board organized a program, started in the summer of 1987, to drill to a depth of between 15,000 and 20,000 ft. The American Gas Research Institute is making a significant financial contribution, on the basis that the Swedish experiment will tell us much about questions of the origin of hydrocarbons and, with that, about the global availability of gas in hitherto unsuspected regions.

If hydrocarbons are found in significant quantities at these depths in the Swedish granite, any explanation in terms of biological sources will seem improbable. If large commercial amounts of gas are discovered, then it will suggest that much gas is indeed streaming up from the mantle and that, therefore, many other locations may exist on the Earth which, like the Siljan Ring, were previously not suspected of being hydrocarbon reservoirs. For the reasons we have discussed, gas rather than oil would be expected in large quantities.

The depth interval between 15,000 and 30,000 ft, which is within reach of present-day drilling techniques, but which has been explored only in very limited regions, could then be expected to be very productive, since that is the region where upwelling gases will generally still be arrested. At the shallower levels, such gases can generally escape quickly into the atmosphere, except where a particularly impervious rock serves as a cap. The drilling in the Siljan Ring may, therefore, be of importance both as a scientific experiment, to understand better what fluids may be making their way up from the mantle, but also quite possibly as a guide to the world's energy future.

References

Chekaliuk, E.B. (11976), The thermal stability of hydrocarbon systems in geothermodynamic conditions, in Kropotkin, P.N. (ed.), ***Degazatsiia Zemli i Geotektonika***, pp. 267–272.

Kent, P.E. and Warman, H.R. (1972), An environmental review of the world's richest oil-bearing region – the Middle East. *Internat. Geol. Cong. 24th, Sect. 5*, pp. 142–152.

Mendeleev, D. (1877), L'origine du pétrole. *Revue Scientifique*, **2e** Ser., VIII, 409–416.

Robinson, R. (1963), Duplex origin of petroleum. *Nature* **199**, 113–114.

Robinson, R. (1966), The origins of petroleum. *Nature* **212**, 1291–1295.

CHAPTER 6

Drilling Technology

S. Varnado

6.1. Introduction

If Gold is correct (see Chapter 5), the world is awash in methane. To recover it, we have to drill deep wells. We cannot prove the theory unless we actually drill, and we need new drilling technology to make this resource economically viable.

This chapter describes where drilling technology is today, how it got there, and what the future is likely to hold. My perspective on the future is tempered by today's reality. I live every day in a profit-and-loss world, which is most often a loss-world today, and times are hard. We have all looked into the future, and it looks promising, but as an industry we must survive the next few years in order to enjoy those prospects.

6.2. Problems of Hard-Rock Drilling

According to Gold's theory, gas will be found in hard-rock areas that are deep and hot. This is a hostile environment for drilling, but we are accustomed to drilling in hostile environments. We are now drilling in 2,000–3,000 feet of water offshore in the Gulf of Mexico. We are drilling in the North Sea. Both of these are hostile areas; but the problems differ as the depth, hardness, and temperature of the rock increase.

In the USA, we have historically drilled at shallow depths. *Figure 6.1* shows 1980 statistics, but the data as of 1985 do not differ significantly. The median depth in US land drilling is about 3,300 feet; the average depth is a little over 4,000! Gas wells have traditionally gone a little deeper: about 5500 feet on average (*Figure 6.2*). There have been some wells drilled in the 20,000 – 30,000 foot range, but not very many. Most of our drilling practices are tuned for shallow land drilling.

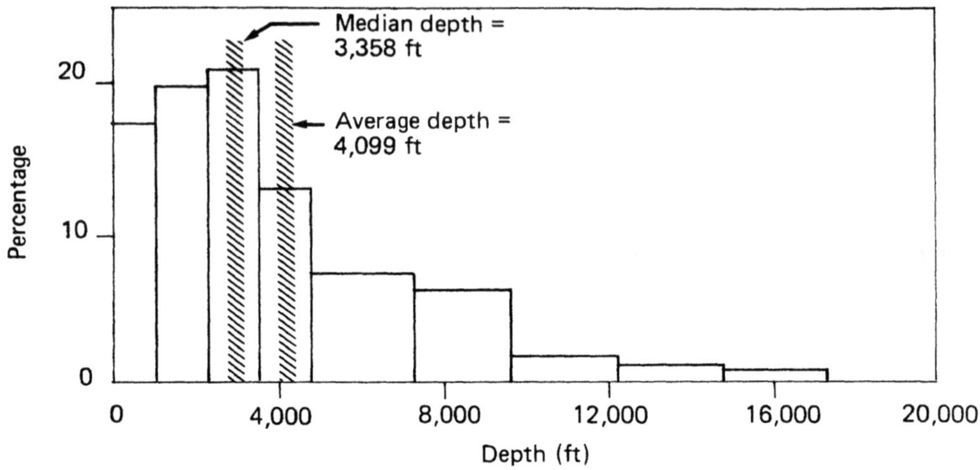

Figure 6.1. US drilling depth, 1980. (Courtesy NL Sperry-Sun.)

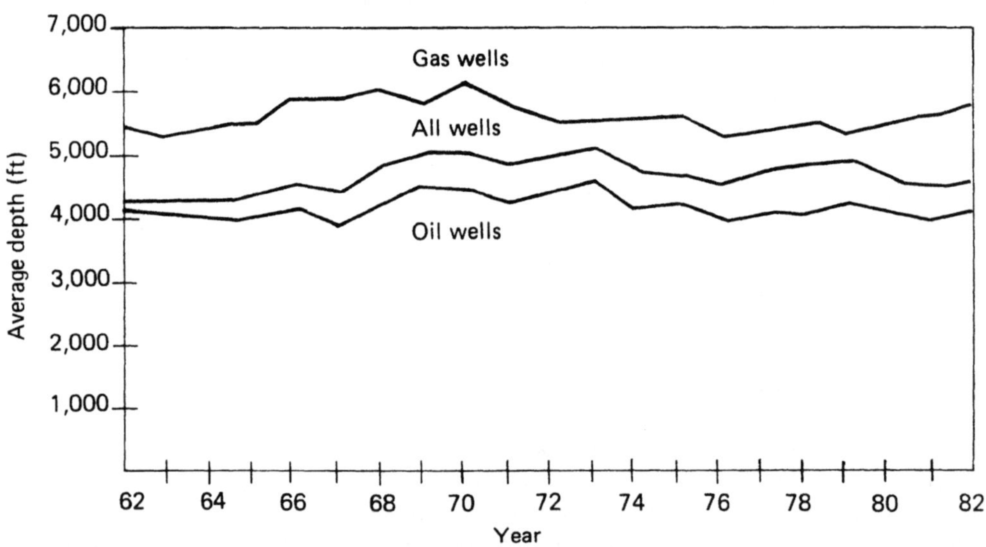

Figure 6.2. Average well depths, 1962–1982. (Courtesy NL Sperry-Sun.)

Drilling costs increase dramatically with depth. Deep rock is almost always harder, more abrasive, and more highly fractured. Rock fractures can lead to a phenomenon called lost circulation: when mud is pumped down to clean the hole, it may not return to the surface, and cuttings are lost in the fractures. This is a difficult problem in geothermal wells and in fractured reservoirs.

High temperature also presents problems. Most drilling equipment is useful to about 300° F. Above that temperature, drilling mud dries out, electronics do not work, and rubber seals and lubricants are ineffective. In addition, erosion and corrosion problems increase with depth. H_2S and CO_2 are usually present. This is a very severe environment for drill pipe. Erosion problems arise if aerated fluids are used to circumvent lost circulation.

These are the problems that we must address if we are going to bring about the "Methane Age".

6.3. Hardware Trends

Developments in drilling technology, like computers, fall into two basic categories: hardware and software. Hardware includes bits, downhole tools, fluids, and completion technology. We also have some new systems, called measurement-while-drilling (MWD), that allow the well to be logged as it is being drilled. In the software area, there is new emphasis on drilling simulators and artificial intelligence, which are expected to take great strides in the next five to seven years.

The first hardware problem encountered is in the drilling bit itself. For example, a bit that drilled in a geothermal formation in California, where the downhole temperature was 250° C, drilled about 500 feet in 26 hours. The cutters and the bearings were entirely worn out. In addition, it was half an inch under gauge. When this happens, the hole becomes tapered. Then, when the worn bit is replaced, the new bit is required to ream back to bottom. This takes time, thus increasing costs.

To solve these problems, we tried using new materials in the existing bit designs. Materials were hardened, bearings were improved, and new lubricants and seals were developed. We achieved a roughly 20 to 30% improvement.

A promising new bit design, which will eliminate some of the problems with roller cone bits, uses synthetic diamond cutters, called polycrystalline diamond compacts (PDC). The disk-shaped diamond layer is brazed to a tungsten carbide post. At first, it was difficult to get the cutters to stay on the bit when drilling hard rock. The brazed bonds simply were not strong enough to withstand the environment. General Electric improved this brazing technique and solved the problem. Another problem is that the cutters fail at about 750° C. Temperatures can exceed that level when drilling hard rock, owing to friction. Careful design of bit hydraulics and cutter placement is required to achieve any degree of success in these applications.

These cutters can be configured in many different shapes. One of the first bits that was built for geothermal drilling applications was designed by Sandia National Laboratories and built by General Electric. There are no moving parts. Cutter positions can easily be changed to match drilling conditions. Tested in the Baca location in New Mexico in 1978, the drilling rate with the new bit was four times faster than had ever been achieved in this geothermal formation.

Moreover, there are many different ways to arrange the cutters in terms of their hydraulics pattern and cutter placements. Research continues to determine optimum designs. Substantial progress is also being made now in understanding the cutter–rock interaction. PDC bits are being run extensively in the North Sea. They work very well with oil-based mud. They are also being run in the Gulf of Mexico. Recently, one bit drilled 6,000 feet without stopping. When removed from the hole, it was still in such good condition that it was used to drill two more wells. Thus, it can be seen that PDC bits provide long life and high penetration rates; however, they also utilize higher mud flow rates. This affects the total system design for the drilling operation. Good systems engineering practices must be maintained to optimize the entire system, especially as the depth and costs of the wells increase.

One way to use the PDC bits is with downhole motors. Several companies manufacture turbines for downhole use, and the Russians have used this technique extensively. Here, the drilling mud provides the hydraulic power required to turn the turbine blades. These systems turn the bit at 800–1,000 rpm rather than the 100 rpm common in conventional rotary drilling. The main problem in high-temperature applications is that standard rubber seals can be used only up to about 300° F. Recently, PDC diamond thrust bearings, which can be lubricated by the mud flow, have eliminated the need for seals, thus allowing operation at higher temperatures.

The use of downhole motors with PDC bits allows operation at low bit weight while still maintaining high penetration rates. This in turn reduces the tendency of the bit to deviate in hard rock.

The next problem to be solved will be the drilling fluid itself. Conventional muds can be used in the shallow part of the hole, up to about 200–300° F; but at a certain depth, a high-temperature mud will be needed. Much work has been done on instrumentation to characterize muds at high temperature and high pressure, and we have a facility that can test muds up to 20,000 psi and about 700° F. Much work has also been done on clay morphology to understand what happens when the clay is heated. Fluids are supposed to be circulated, but at high temperatures conventional muds can cake onto the stabilizer. Sometimes cooling towers are used on the surface to keep the mud temperature below that at which it cakes up.

One technique for preventing loss of circulation is to reduce the mud weight. In many cases foam can be used. Foam has very good lifting capability for removing cuttings, but there are some practical problems in working with it. In drilling applications, the soap molecules start to break down at higher temperatures. In testing done at the Drilling Research Laboratory in Salt Lake City by Sandia National Laboratories, a specially formulated foam was heated to 400° F and maintained its structure for a day. It is likely that foams can be formulated that will be stable up to 600° F.

Several options are, therefore, available for drilling fractured formations at high temperature. A good deal of work has been done on high-temperature muds, high-temperature foams, air drilling, and nitrogen drilling. The point is, however, that careful planning is required for successful drilling of these formations. I think a good well plan should provide for contingencies, because if unanticipated problems arise, the solution may be very expensive.

Deep wells in hot, fractured formations will be difficult to complete (*Figure 6.3*). Setting casing in fractured formations is sometimes difficult. The cement goes into the fractures, leaving voids behind the pipe. If the fluid produced by the well is hot, the casing will expand and sometimes collapse. There have been cases in geothermal wells where the wellhead came out of the ground about seven feet when steam production was turned on. The point is that standard oilfield technology may not be good enough for some of these deep gas wells. We must make a distinction between technology for oil wells and technology for gas wells.

The cement itself is also a problem. At high temperature it may set up prematurely. This leads to improper placement and unsupported pipe. Casing collapse is not uncommon.

Another problem is getting the cement to displace the drilling fluid. At high temperature, the mud tends to set up in the hole. If it is not displaced by the cement, the danger of casing collapse arises again.

In geothermal wells, retardants are used to keep cement from gelling at temperature. If too much retardant is used, the cement never hardens. The key is good planning.

Finally, little is known about the stability of boreholes in the type of rock found in the Siljan Crater. Will the rock spall? Will it slough off and fill the hole? Will it have to be cleaned out? These are some of the unanswered questions about deep boreholes.

6.4. Information Technology

One new development receiving a lot of attention is measurement-while-drilling (MWD) equipment. Basically, the idea is to get some of the logging sensors near the bit, so that measurements can be made before mud invasion occurs. In most cases, wireline logs are not available until two or three days after the well is drilled. The mud sits in the hole, invades the formation, and it can mask the true readings of rock resistivity. With MWD equipment, sensors can be placed directly on top of the bit. These data can be telemetered up to the surface using the mud column. The pressure in the mud is modulated by a downhole valve. At the surface a pressure transducer is used to extract that information. Logs of gamma ray, resistivity, and porosity can be plotted. Directional information is also provided in real time. This is very important for maintaining directional control when drilling in hard rock. These sensors operate up to temperatures of about 280° F.

To generate the mud pulse, we cut a small hole in the side of the drilling collar. A shear valve is placed in this hole. The shear valve is modulated by signals from the sensors, usually once per foot. If the data are not transmitted to the surface in real time, they can be recorded downhole and

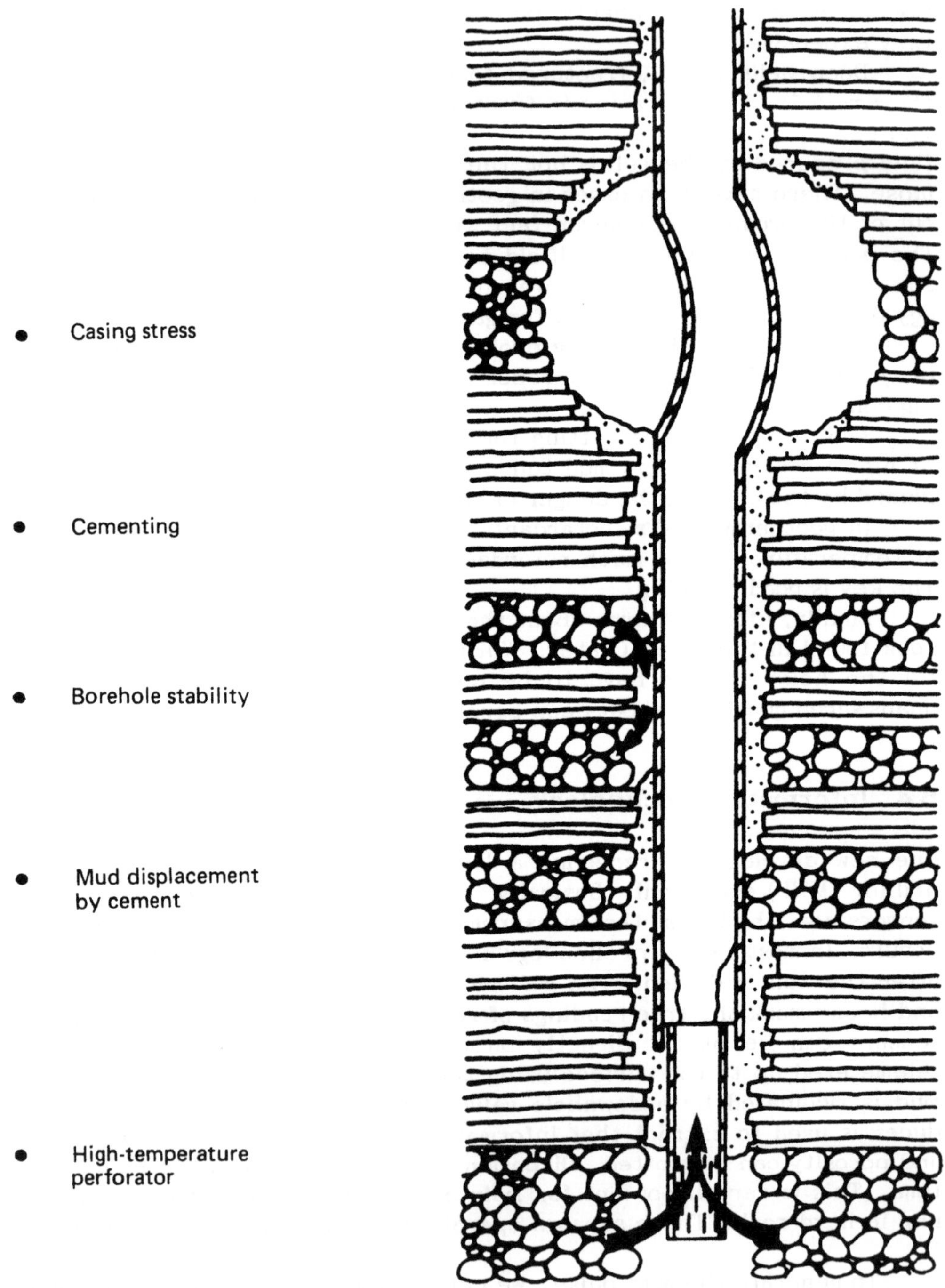

Figure 6.3. Completion technology problems in hot, fractured formation. (Courtesy NL Sperry-Sun.)

recovered when the tool comes out of the hole to change the bit. About 300 hours of data can be stored.

There is a new resistivity sensor called Electromagnetic Wave Resistivity (EWR) – service mark of NL Industries. This sensor has a resolution that is comparable to a medium induction log. As a matter of fact, this tool can replace many conventional wireline logging runs. In an offshore well, logging costs may be as high as $700,000. With MWD, the costs are much less. MWD also aids the drilling process. It allows accurate pore pressure prediction and selection of casing points. There are big changes under way in the entire wireline logging industry, and this tool is going to be one of the ones that make a difference.

The mud pulse valve works by opening for a short period of time and then closing. A small volume of fluid is vented to the annulus when the valve opens. This is called a negative pulse system. The data rate for this system is about 3 Hertz.

Figure 6.4 diagrams a very high data rate drilling telemetry system developed by NL Industries. This system has a wire from the surface down to the sensors. It has been the dream of drilling engineers for 30 years to know exactly what that bit is doing. Shell tried to do this in the 1970s, GE tried to do it with Electro-drill, and Exxon is trying to do it now. Those earlier efforts were moderately successful. A cable is placed inside the drill pipe. A section, say, 1,500 feet, of prewired drill pipe is used while drilling ahead. There is an electrical connector in each tool joint. At the surface, slip rings are used in a special kelly to allow electrical power and data to be sent downhole.

NL Industries' system was tested last summer. The near-bit mechanics sensor in this tool measures the bending moment, the borehole pressure, the acceleration in three dimensions, the temperature, and the weight-on-bit and torques (*Figure 6.5*). These data were transmitted to the surface at 20 kHz. This is the first published mention of these high data rates.

This test was done in a straight hole, about 8,000 feet deep. There was an indicated weight-on-the-bit of 30,000 pounds at the surface. *Figure 6.6* shows what was really happening at the bit as a function of time. The weight-on-the-bit was increasing to 70,000 pounds and decreasing to zero. The bit was actually coming off bottom, but the surface indication was 30,000 pounds on the bit, and everything seemed fine. What is actually happening is that the bit is bouncing. In hard rock this is a severe problem because it reduces the lifetime of the cutters. The torque at the bit is also shown in this figure. Note that torque reversals are occurring. This phenomenon causes accelerated bearing wear in roller bits and causes cutter damage in polycrystalline diamond bits.

This type of data can be used to analyze drill string dynamics and to design shock absorbers. Currently, shock absorber design is not well advanced. Until now, data on the forcing and response functions have not been available. It should be mentioned that this system is not yet available as a commercial product – currently, it is too expensive. NL, Shell, and Standard conducted a joint field test to get these results. The data are now

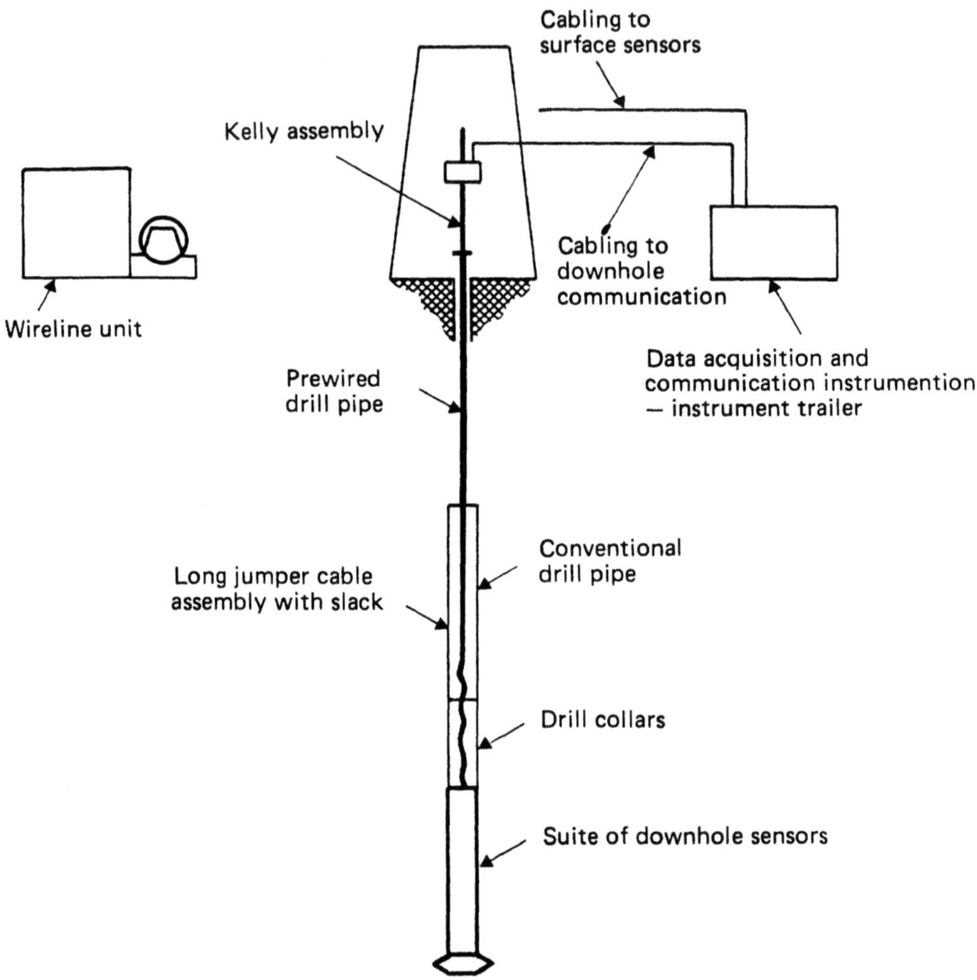

Figure 6.4. Principal elements of drilling telemetry system developed by NL Industries. (Courtesy NL Industries.)

being analyzed, and it will take a long time to bring this system into commercial use.

The bending moments that we measured in this hole were about ten times the values expected from static analyses (*Figure 6.7*). If any H_2S is present with these high bending moments, stress-corrosion cracking accelerates tremendously. Pipe will wear out quickly. Changing the rpm and weight-on-bit can reduce the bending moment. Care must be taken in selecting the drilling parameters to achieve optimal performance.

MWD is one source of data. For system optimization, all available data should be integrated and utilized. It is now possible to make measurements on the surface and transmit this information back to the customer's office.

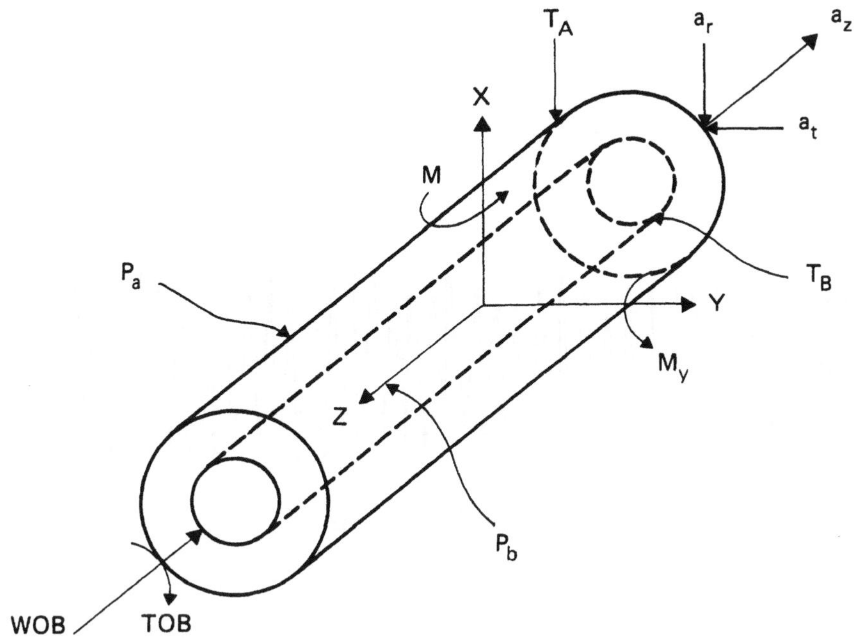

Figure 6.5. Measurements made by near-bit mechanics sensor. (Courtesy NL Industries.)

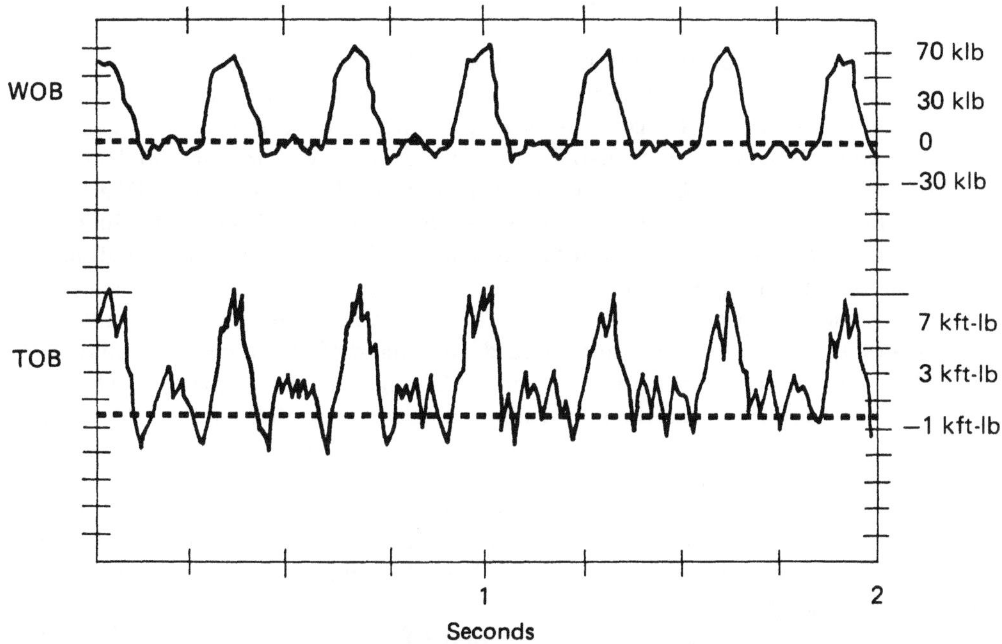

Figure 6.6. Weight-on-the-bit (WOB) and torque while bouncing off bottom at 70 rpm and 30 klb. (Courtesy NL Industries.)

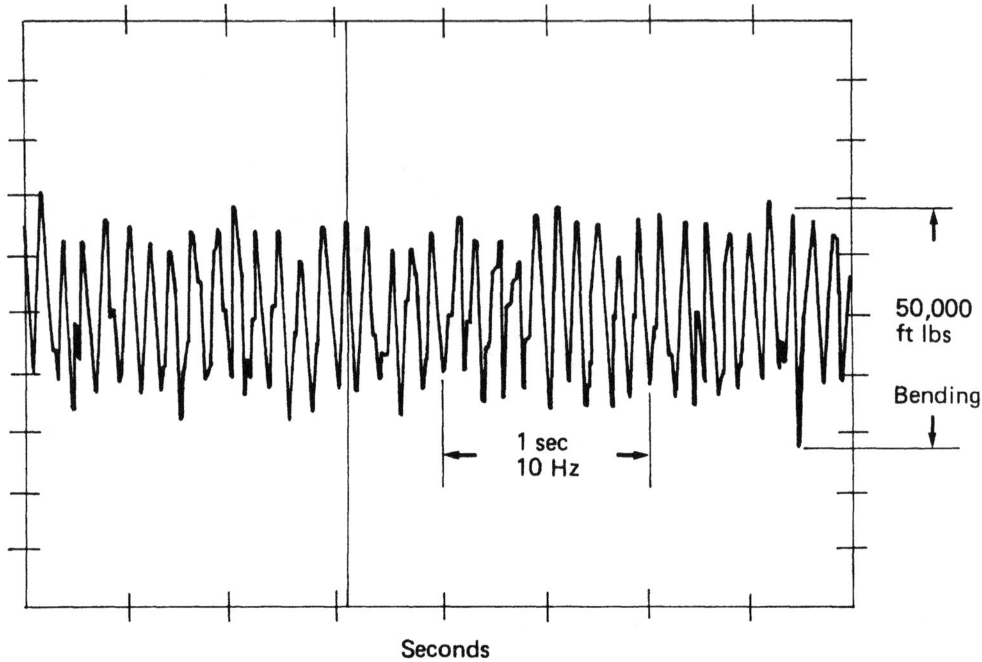

Figure 6.7. Bending signal in time domain at 120 rpm; BHA 4A without undergauge stablizer. (Courtesy NL Industries.)

A useful new concept is that of a rig site data center, in which data from surface sensors, the MWD systems, survey information, etc., could be collected (*Figure 6.8*). Interpretive models could also be run on the rig. Data storage and displays would also be provided so that the man on the rig is looking at the total system. This information can then be transmitted into a central data center in town, where expanded data monitoring and interpretation facilities would be available. Normally, CRT displays are available in the data center along with hard copy.

In a small data center NL set up for Tenneco, operators are on duty 24 hours a day, seven days a week. The data from different rigs are displayed on CRTs, and, at the present time, up to 20 rigs can be served by this data center. An off-line processor manages the graphics and interpretive software. With the data in the office, the man on the rig and the man in the office have exactly the same information. If there is a problem on the rig, help is available. With a terminal in his office, the engineer can dial up the rig and learn exactly what is going on in real time. The real-time data are used to improve efficiency, monitor the drilling operation, and provide technical support when problems arise.

Figure 6.9 measures the time it takes for the crew to add a length of pipe. Normal procedure is to drill 30 feet, pick up another 30-foot joint of pipe, and screw it into the drill string. This is called "making connection".

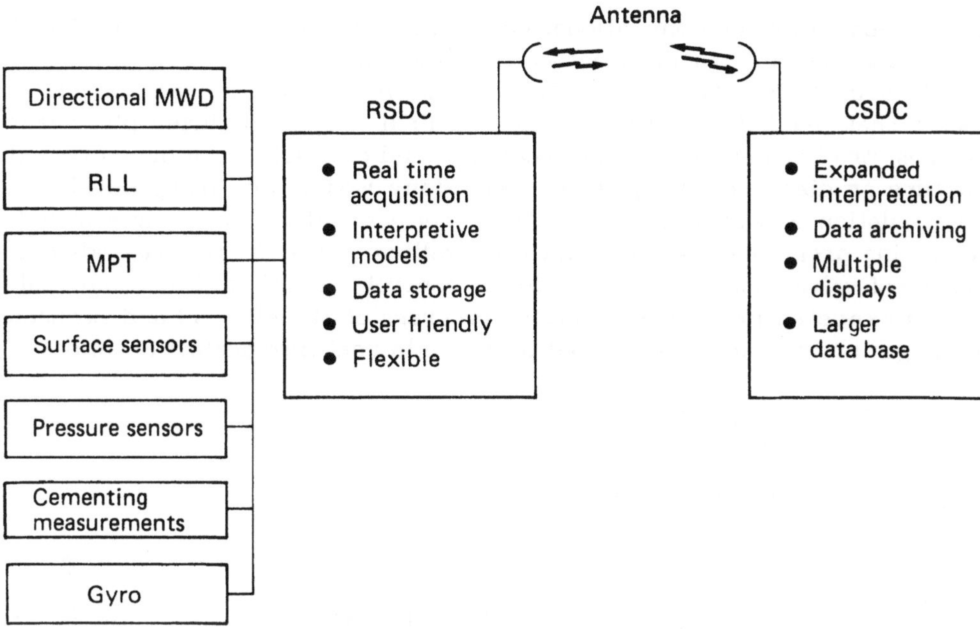

Figure 6.8. Integrated drilling information system concept.

This may seem to be a trivial thing – it only takes about a minute. However, there are hundreds of connections made during a bit trip. From this figure one can see that some contractors were taking more time than others. Awareness of these differences leads to improved efficiency.

By actual measurement, Tenneco saved 17.5 hours per well after they started monitoring the connection time. This savings in time translated into about $40,000 a month, just from providing better management.

Data that are monitored on a routine basis are as follows:

- Hookload.
- Depth.
- Rate of penetration.
- Torque.
- Mud weight.
- Mud temperature (in and out).
- Mud flow rate (in and out).
- Pump pressure.
- Choke pressure.
- Pit volume.
- Gas units.
- Mud conductivity.

The data are updated every 10 seconds. MWD measurements can also be transmitted into the data center. Gamma ray and resistivity data can then be displayed in the office while the well is being drilled. This is a marked improvement over the conventional way of doing things, where the geologist might wait three or four days before seeing log information.

Interpretive software is needed to make full use of the real-time data. Models that predict the direction of the drill bit as stabilizer placement changes, predict invasion and permeability, and calculate pore pressure and fracture gradient are being developed. Many models are being developed for simulation and system optimization. For example, there exists a kick monitoring program of gas flows into the wellbore: the driller must adjust a number of valves and fluid properties to control the influx. This model calculates expected pressures, monitors progress of the kill mud and position of the gas bubble, and plots actual pressure in real time against anticipated

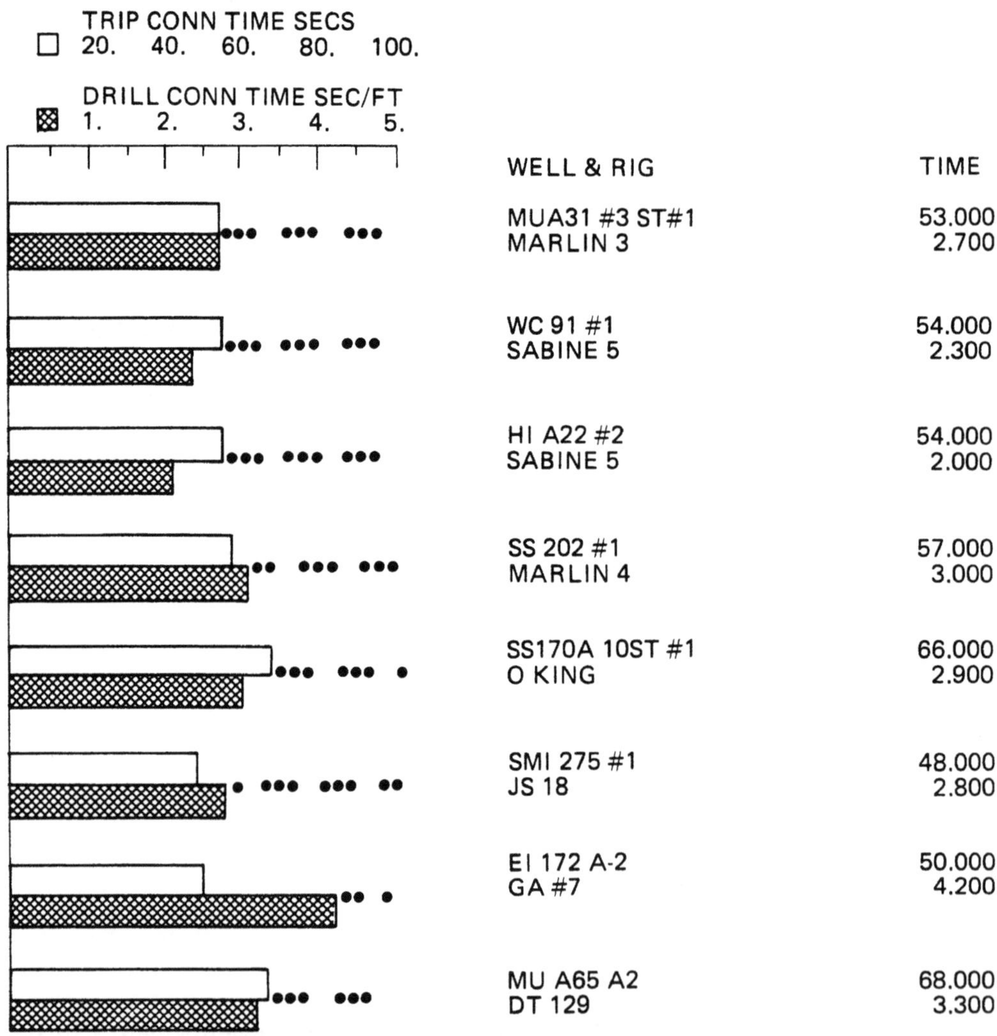

Figure 6.9. Tenneco drilling and trip connection comparison for March 1985. (Courtesy Tenneco and NL Baroid Logging Systems.)

pressure. The benefit of this is that the man out on the rig has help available in the office, monitoring the operation and making suggestions.

6.5. Anticipated Impacts of New Technology

Technology has advanced, and we understand the drilling operation better now than ever before. Can this technology be effectively utilized in the field?

To answer this question, we have to look at what is happening today. Things are dismal in the oil field. *Figure 6.10* shows what happened to the US rig count during the past few years. In 1981, there were 4,500 rigs operating in the USA. In 1983, we started out with about 2,600 rigs operating, dipped down in April–May to below 2,000, but then recovered. In 1984, the same pattern occurred. We thought 1983 and 1984 were terrible. In 1985, the downturn really started. The count went down, and it stayed down. We did not get the third and fourth quarter upturn that everybody expected. In the first five months of 1986, we dropped from 1,900 rigs down to 828 rigs. In the Gulf of Mexico we have about 100 rigs running, where there were 232 in December 1985; 38% of the fleet is working – the other 62% is stacked.

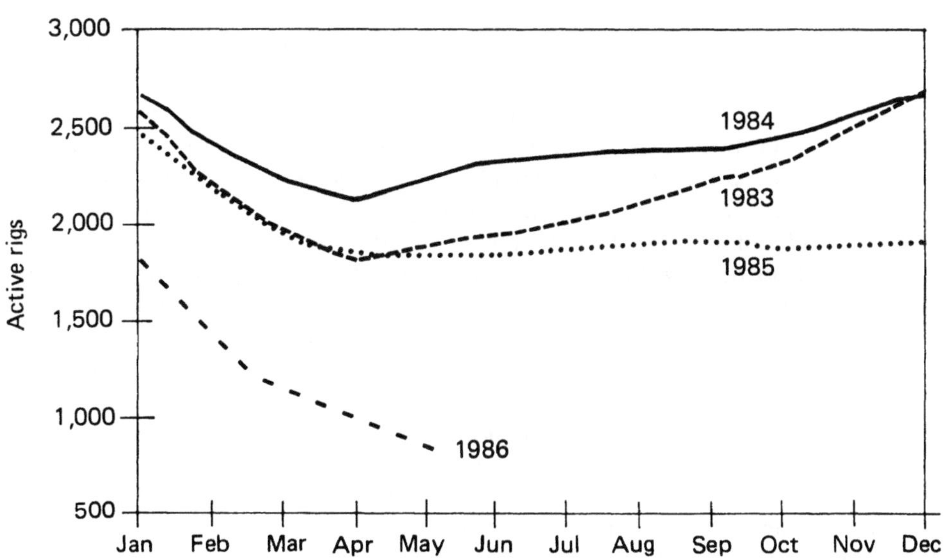

Figure 6.10. Total US rig count, 1983–1986.

In short, the rig count dropped from 4,500 rigs down to 800 in four years. I do not know of any other business that has seen that kind of decline in such a short period of time. It has been tough, and for the last two or three years drilling contractors, service companies, and operators have been cutting back, laying off employees, and trying to figure out how to survive. These conditions have created low morale throughout the industry.

This downturn has happened so quickly and gone so far that structural changes begun in the industry in 1982 are now being accelerated. When we come out of this tailspin, assuming that we do, the oil and gas industry will be different. Prior to 1982, business was done in the conventional way. During that time demand typically exceeded supply. We were growing, we needed more oil, and prices were high for services and for the product. We had undercapacity – not enough rigs, not enough drilling tools – and we had a very fragmented supply industry. We bought bits from bit manufacturers, mud from mud suppliers, etc. Somebody had to put all that together and make a drilling system. This normally fell to the company man on the rig. There were a lot of different suppliers, primarily using US technology. I believe that the USA led in drilling technology up to 1982.

There exists, however, a lack of global perspective in the management of many of the companies in our industry. Few of the people managing the service companies were able to recognize the possibility of natural gas becoming so important. We did not understand the Middle East situation, so we were always in a reactive mode. When we saw the downturn in 1982, we started cutting back a little bit, but nobody expected the total collapse that has occurred.

I believe what is happening now is that we are in a kind of parenthesis in the industry – a pause. In 1982, supply shot up past demand. We have tremendous uncertainty in prices today. You can poll any group of experts, and you will get different answers for future oil price scenarios. Some will say it is going back $20; if so, things will improve. I say the price might stay at $15 or go back down to $10. The longer it stays down, the quicker it is going to come back: it may go to $40, if it stays at $10 for a year. In any case, during this interim period, there is great uncertainty. We have overcapacity, particularly in the service industry. A friend of mine bought a 30,000-foot drilling rig two weeks ago. It cost $9 million new: he paid $300,000 for it, and it is in great shape. That is what overcapacity means: the bank owned the rig and was glad to get rid of it.

There is going to be a massive restructuring and down-sizing in the service industry and in the oil companies. They are going to do things differently. There is not merely poor morale, but chaos in this industry at the moment. The leadership is up for grabs. We have perhaps 20 service companies now; I predict that only four or five may survive this change. I think that, by 1990 or so, we will have a new industry.

I believe Tom Gold (see Chapter 5) is right. I think we are going to drill deeply and find natural gas. There is going to be an emphasis on technology. We will not drill these wells cheaply without using the best available technology. There will be a return to the systems approach. The winners will be the people who can look at drilling as a system.

There will be fewer suppliers. You will see more packaged services, i.e., these fewer suppliers will provide all the services required on a well. This has already started. It is part of the massive restructuring. We are seeing mergers and acquisitions. Unfortunately, from the US perspective, there seems to be a shift in technology development and leadership. Technological leadership may shift to the North Sea region. The Norwegians and the British are accelerating technology development. Schlumberger's Cambridge Research Center is going to turn out some good work in the next three to five years. The decline in the US industry is going to be offset by increased technology development in other countries. I think there will also be new financiers for gas technology. I am not sure the oil companies are going to step in and finance Gold's ideas. I think you will see money coming in from other people who are not as conventional, not as bound by the old tradition.

What is going to happen in exploration? In the 1980s, some of the fastest and best computers in the world were used to aid geophysicists. Three-dimensional seismic analysis became popular. I fear that there was too much dependence on the computer results and not enough systems thinking about where the resources might be found. Few people examined all the available data and integrated it to make decisions. There were no major finds in the 1980s. People who found reserves were those who relied on good, thoughtful geologists, not big computers.

In the future, the integrated approach to thinking about what we are doing will resurface. We have to view exploration as a system. That is why I am optimistic about what Gold and IIASA are doing. They are looking at exploration as a system. They are developing an understanding of why the gas should be there before drilling for it. The same thing will happen throughout exploration. We will sit down as systems thinkers and figure out where the reserves may be. We are going to use all available geophysical data, but there will also be more emphasis on drilling and coring. We will have new technology for coring. We are going to drill more holes with continuous coring rigs using low-torque drill pipe. The rotary speed will be about 600 rpm. It will be a computer-controlled rig with robotic pipe handling. We will drill these exploration holes at one third of what they cost today. Then exploration and confirmation holes will be drilled to make sure we are right. This is the way to expand exploration budgets and make the most out of limited financial resources.

What is happening now in drilling technology development? First of all, budgets are being cut. Only one US operator has an active drilling R&D program, and that is Amoco. The others have small groups, but they are focused on short-term problems. Outside the USA, the Norwegians are starting up a drilling research institute and Schlumberger's Cambridge Research Center is going to be a world leader in drilling research. That is why I think leadership may shift to the North Sea.

In the USA, technical people are leaving the industry. Mechanical and electrical engineers who can leave are probably going to do so. We are losing good people right now in the USA to other industries. There is a reduction in training programs. University enrollment in petroleum engineering is dropping rapidly. Petroleum engineering schools in the United States are going to be severely crippled. This is a sad thing to see. It seems we are essentially dismantling the US industry. If the situation is not reversed soon, shortages of both trained people and equipment will result.

So what is going to happen? I think we are talking about a brand new industry, totally different than it was before 1982. There will be more joint venture projects, e.g., service companies getting together with drilling contractors to do joint research. Systems engineering skills will be in strong demand in the future, but technical leadership, as I said, is going to shift from the USA to the North Sea. You are going to see the use of high-fidelity simulators for training, planning, and drilling optimization. This is the only way that systems integration can effectively take place. Use of telecommunication will increase. The new research areas are going to be artificial intelligence, smart sensors, and automated, robot rigs.

What do I mean by smart sensors? If we are going to use data bases, models, and simulation, we have to ensure that the quality of the data is very high. To accomplish that, some intelligence is needed in the sensor to make sure it is working properly. It will be self-calibrating. To date, however, this is an infant technology. Little progress has been made on drilling sensors in the last 15 years. We still measure density by weighing the mud.

Who is going to survive? Companies that are entrepreneurial, that are willing to take some risk, that go out and try to make something happen. The survivors will have to have technological know-how and be willing to make the changes required. They must have bifocal vision, i.e., they must look at short-term cash flows versus long-term market position. Balancing these perspectives is a difficult management problem. Companies that cannot do it are going to lose. Successful companies will have a systems orientation. The key to it is people. If we can hold onto the people who have the skill and the spirit – right now the spirit is what is lacking – to effect the turnaround, then when you get ready to drill for all this natural gas, we will be in a position to do so.

CHAPTER 7

Methane Technology: A Technical Survey

Rodney E. Hanneman

7.1. Introduction

The discovery that the continents have moved laterally over great distances has led to major breakthroughs in our understanding of the geodynamic processes shaping the earth's landscape in recent years. Yet the concept of a mobile Earth has been the subject of controversy since the eighteenth century.

In 1915, Wegener published a particularly stimulating article in which he claimed the lateral migration of continents provided a mechanism for the process of mountain building. This work had a great impact, but it required another half century before the revolutionary concept received widespread acceptance in the general scientific community.

Again, we are faced with revolutionary concepts about potential major natural gas sources, presented by Gold (Chapter 5) and others. If history repeats itself, broad acceptance of some of these concepts will take longer than many would like. Through this transitional period, as is true with almost any major change in geological thinking, many incremental modifications will be made as new data emerge. This chapter presents several different perspectives on abiogenic gas, as well as an overview from the Gas Research Institute and industrial perspectives.

7.2. Role of the Gas Research Institute

Natural gas is basically a clean fuel, requiring lower maintenance for many systems. It carries relatively low environmental risk and is relatively easy to produce. In the USA, the domestic infrastructure is in place with plenty of excess capacity in many areas. People have more or less accepted the safety of transport. However, the key question we face is not whether

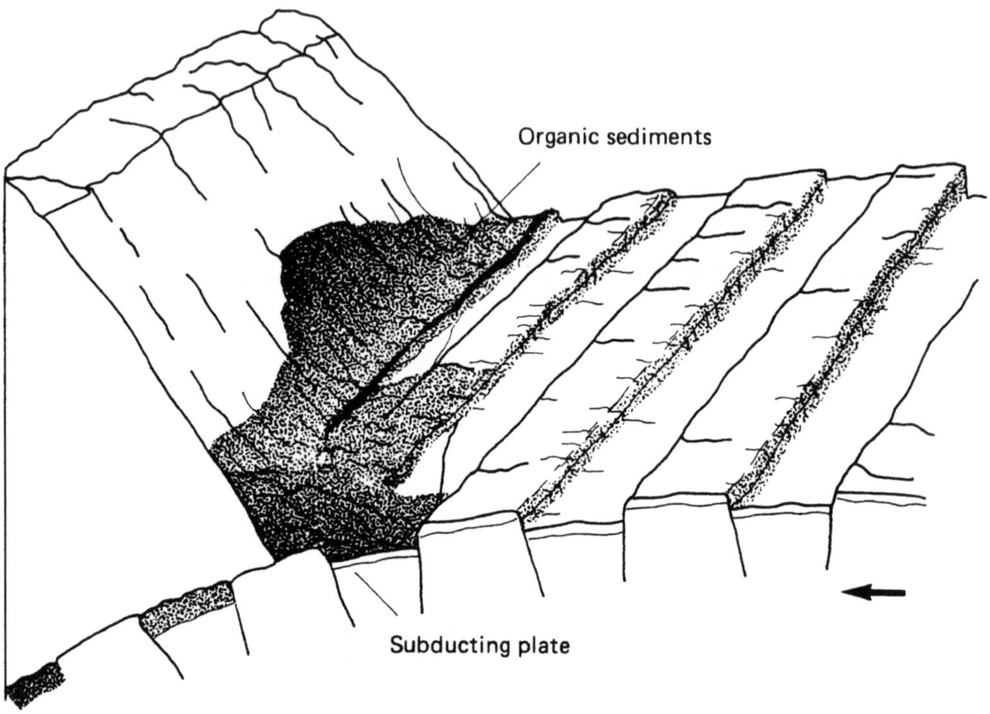

Figure 7.1. Sediment subduction at the continental margins (Hilde, 1984).

there is adequate natural gas, in terms of cumulative quads in the ground, but rather what the cost is going to be to recover it.

In the USA, we had an overabundant supply of gas, which was originally driven by gas associated with oil deposits. People became accustomed to getting gas at abnormally low costs, and this set in motion a train of circumstances that gave very little incentive for strong technology development by the gas industry. It was only after the first OPEC oil crisis and related aftershocks that the US gas industry was compelled to invest in improved technology, if the industry was to survive in its own right – and that is when GRI was founded, approximately ten years ago.

My particular role in GRI at the moment is Chairman of the Research Coordination Council, which advises GRI on technical matters. I briefly outline below GRI's program and nature, and then discuss some of the more specific technology and cost tradeoff issues for various energy sources.

GRI is a quasi-private (or quasi-public) nonprofit institution. Currently, its total budget is about $187 million, most of which is project budgeted ($157 million). Currently, the funding split allocates 20% to supply issues; 54% for end-use research (some of which is generating totally new end-uses and some of which is energy conservation-oriented, aimed at making gas more competitive with other energy sources); and 12% is spent on operations work, which is largely related to safety and transmission; the small balance goes to what we describe as cross-cutting research. The geoscientific portion of the budget is very active at present, covering everything from gas sampling through field studies. This has been an evolving program over the last several years. The Siljan Crater experiment, for example, is budgeted at GRI for about $4.3 million. The Swedish government will cover approximately 80% of the total experiment cost.

The current outlook is that the USA has a gas bubble. It is hard to move gas these days at any price if you do not already have some contracts lined up to take it. This is the combined result of heavy drilling in the late 1970s up through 1981, when there were as many as 4,500 rigs running. The very dramatic recent oil price decreases have been a tremendous shock to people, in both the domestic oil and gas business.

Basically, the situation since 1981 has destabilized our drilling business. Many companies are going out of business. Many geologists are moving to other lines of work. If prices stabilize in the $10 or $12 per barrel range, permanent damage will be done to our infrastructure, which, when the bubble disappears, could make the first oil OPEC crisis look like a picnic. Thus, we hope to see some rational balance returning to the pricing structure. Meanwhile, GRI is working aggressively in terms of supply orientation, as well as generating some new end-use businesses.

7.3. Biogenic versus Nonbiogenic Gas Sources

My belief is that both major biogenic and nonbiogenic gas sources exist. The real question throughout the supply-oriented community in the USA – and probably most of the world – is the extent to which nonbiogenic sources exist at a recovery cost that we will find tolerable when compared with other low-cost energy sources. The current belief among the majority of technical people is that, for the short term, there is a rather significant supply of biogenic gas.

Now let me touch on a few intriguing technical facts relevant to abiogenic gas. I had occasion nearly a decade ago to be involved in some work on diamonds somewhat similar to the work of Giardini, Melton, Dienes, and others. Natural diamond single crystals with gaseous inclusions from diamond pipes were fractured in a mass spectrometer, and the gases were analyzed. The composition results give some very important insights as to conditions existing at certain depths in the earth. I think gaseous diamond inclusions are possibly the best single *in situ*, accessible probe of deep geologic chemical conditions, including oxygen fugacities and carbon and hydrogen activities at these particular depths. This probe evaluated conditions far below the bottom of the deepest well, even those in west Oklahoma or the USSR. The diamond–graphite stability line is well known. The gaseous bubbles within the single crystal diamond were trapped during the growth of the diamond deep within the earth.

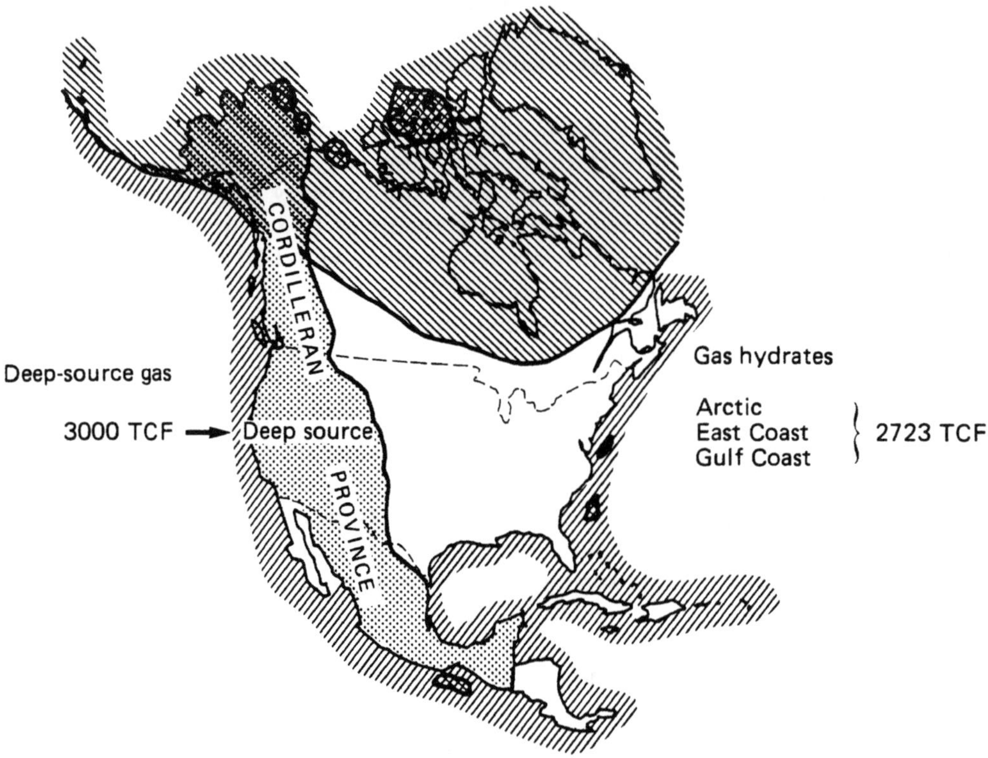

Figure 7.2. Deep-source and gas hydrates of North America: Potential supply after the year 2000.

One critical experiment showed a composition of, roughly, 8% methane and 20% CO_2. Computer models were run on multicomponent thermodynamic equilibration; and from this particular equation, we were able to establish the approximate depth and temperature of bubble entrapment. With certain reasonable assumptions, such modeling shows that methane was at least 30,000 psi at a depth of the order of 50 miles. Thus, there was almost certainly a high partial pressure of methane at that depth level, which is far below any conventional sedimentary basins. However, there was an even higher partial pressure of CO_2 at that depth, where it was entrapped.

One other source of carbon or possibly even hydrocarbon is worthy of mention: various carbon-bearing meteorites, including iron–nickel based ones with graphitic inclusions (such as the Canyon Diablo meteorite), and carbonaceous chondrites, which bombarded the earth violently over many eons.

7.4. The Cost Factor

The largest single determinant of the supply cost is the drilling and completion cost. We read in Chapter 6 how these costs can be reduced in both evolutionary and revolutionary ways. I am of the opinion that the cost of drilling could go down by at least a factor of two, assuming the R&D resources are put to work to fulfill the technologies we can envision.

A second point is that the statistical hit rate on wildcatting for a commercial find is perhaps one well in seven or eight over the past several years. Thus, there is a potential large economic benefit for efficiency in drilling exploratory holes. Obviously, we are not doing as good a job in the USA as in some other places, based on the data presented by Dr. Marchetti (Chapter 3). In any case, an increase in the hit rate alone, through enhanced exploration methods and analyses, would decrease the cost of gas rather substantially. Simulation techniques will be very helpful, both for drilling and exploration.

Finally, we must continuously examine the costs and potential payoffs of finding competing energy resources, such as coal and oil.

7.5. Prospects

There is a range of current and future potential gas options under consideration by GRI and the US Department of Energy. GRI has focused on some aspects of conventional gas in recent years; but tight sands, Devonian shales, and coal seam methane in particular are also being examined.

Methane hydrate, if extracted successfully, could produce 50 cu. ft. of gas per cu. ft. of rock, assuming 30% reservoir rock porosity. Conventionally, we are accustomed to working in reservoirs of sandstone and limestone with lower potential gas yields. When we consider Devonian shales, an order of magnitude less gas per cu. ft. of rock can be expected, coupled with low permeabilities.

Recall that natural gas formations can be classified according to three mechanisms: (1) conventional biogenic origin, (2) reactions of ultra-deep carbon with a prior organic ancestry, and (3) abiogenic gas with no biological history. In the case of conventional biogenic origin, as organic matter becomes buried, a modest percentage of the material is converted to methane by nonthermal methods – namely, anerobic bacteria. This takes place at relatively modest temperatures – in fact, almost ambient temperatures. Carbon dioxide is certainly released in this temperature range, too. Eventually, as buried organic matter is driven to deeper, hotter conditions, it undergoes catagenesis. Various types of hydrocarbons are released, including oils with varying chain lengths, plus some methane. As buried, remaining organic matter is depressed into even hotter, deeper conditions, metagenesis occurs, yielding principally methane and hydrogen sulfide in some cases, in addition to overmatured carbon. Below this depth no more conventional biogenically derived methane is possible. All this conventional, biogenic methane release depends on sedimentation rates, geothermal gradients, tectonic plate events, moving and thinning of plates, and various other factors.

If one considers all the world's known coal beds and assumes they have passed through the normal coalification processes, from peat through anthracite, at their current stages of coalification, and calculates the amount of methane that had to have been released to produce those coal beds, it appears that at least 10,000 times more methane has been released than the most liberal estimates of a resource base of methane today. Though most of that coalification-type methane has been lost by oxidation in the atmosphere, even a few hundredths of 1% of this biogenically sourced methane entrapped in subsurface formations and reservoirs would account for many thousands of quads of resources.

Two primary tectonic-type processes are of interest in natural gas formation – thinning and plate sebduction. There is also a shear-type reaction. According to some experts, perhaps 70% of the biogenically derived natural gas is associated with plate thinning episodes, and 30% may be associated with subduction. Over the past 750 million years or more, there have been dramatic variations in biological activity. Shifting plates versus equatorial positions have also contributed to dramatic variations in buried biological matter around the world.

One prior study of all potential biogenic sources suggested the possibility of several thousand quads of natural gas worldwide at up to $8 to $12/MMBTU recovery costs (in 1980 dollars). While these numbers can only be considered "guesstimates," they do suggest a very sizable remaining biogenic resource.

There is a second class of natural gas – namely, methane potentially produced from ultra-deep subduction of overmaturated carbon chemically reacting with hydroxyl ions and other hydrogen sources deep within the lithosphere and mantle. Consider the many oceanic plates that are colliding with continental crusts. These have a higher density; and, upon collision, there is a subduction zone. Certain organic sediments lay down in these regions, particularly right near the edge. As they are driven down, these deposits go through the sequential thermal maturation processes, releasing various amounts of methane on the way. Eventually, a point is reached, perhaps only 10 kilometers down (maybe 15 or 20), where basically only carbon is left. As this residual carbon (of prior biogenic history) is driven deeper, it reaches the range where oxygen and hydrogen in the mantle rock can chemically react with the carbon (essentially, inorganic carbon at that point). Some thermodynamic equations suggest that this is going on. If so, it is a potentially big source; and it could be occurring in many places. This type of phenomenon is shown in *Figure 7.1*, which is from a US Department of Energy (DOE) study. Over 1,000 miles of deep active subduction zones exist in the various regions associated with the USA alone.

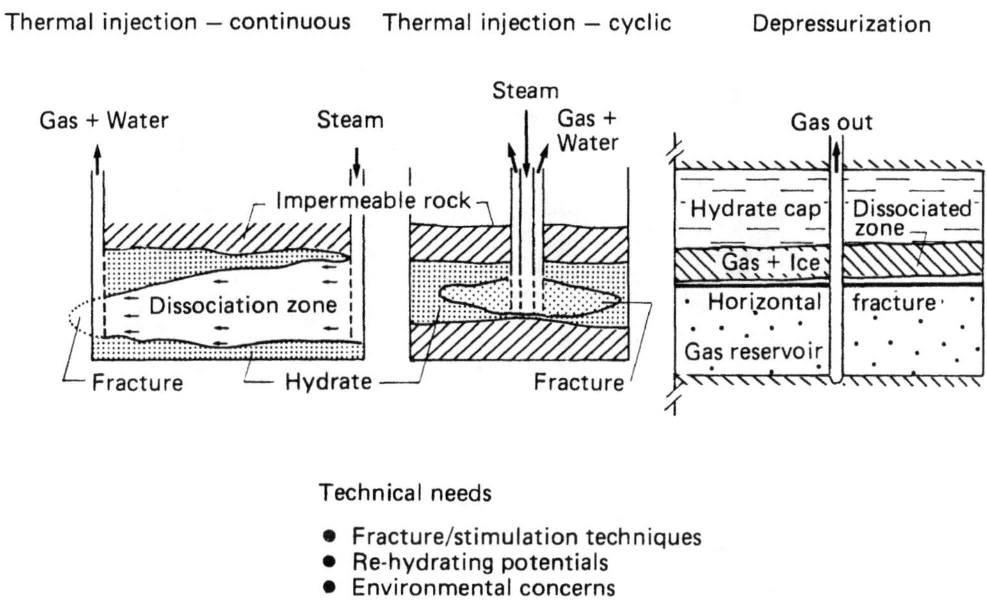

Figure 7.3. Production of gas hydrates: Where free gas is found with hydrates, depressurization recovery schemes work; all other hydrate deposits require thermal recovery schemes.

Vast potential exists for unconventional gas resources in the USA (let alone the rest of the world) according to certain DOE studies, as suggested in *Figures 7.2* and *7.3*. These particular numbers can be considered highly speculative and largely uneconomic with present technology. Potential ideas are also being put forward as to how one might eventually recover methane hydrate.

Tight sands involve a very complex recovery challenge. Such structures consist of various geological lenses, not like typical conventional reservoirs, as shown in at least a few quads; and, in due course, a lot more of them in the USA could become commercially viable – not at today's gas prices, but at gas prices that might rationally come up in the next 10 or 15 years. The technological advances needed to fully exploit tight gas sands are shown in *Figure 7.4*.

A recent coal bed methane assessment indicates that eventually about 400 quads may be potentially recoverable in the USA. DOE and GRI are interacting on various systems to recover some of this coalbed methane.

The abiogenic gas approach described by Professor Gold in Chapter 5 is based on massive accretion and entrapment of methane in the early stages of the earth's history, followed by a long, slow evolution of natural gas.

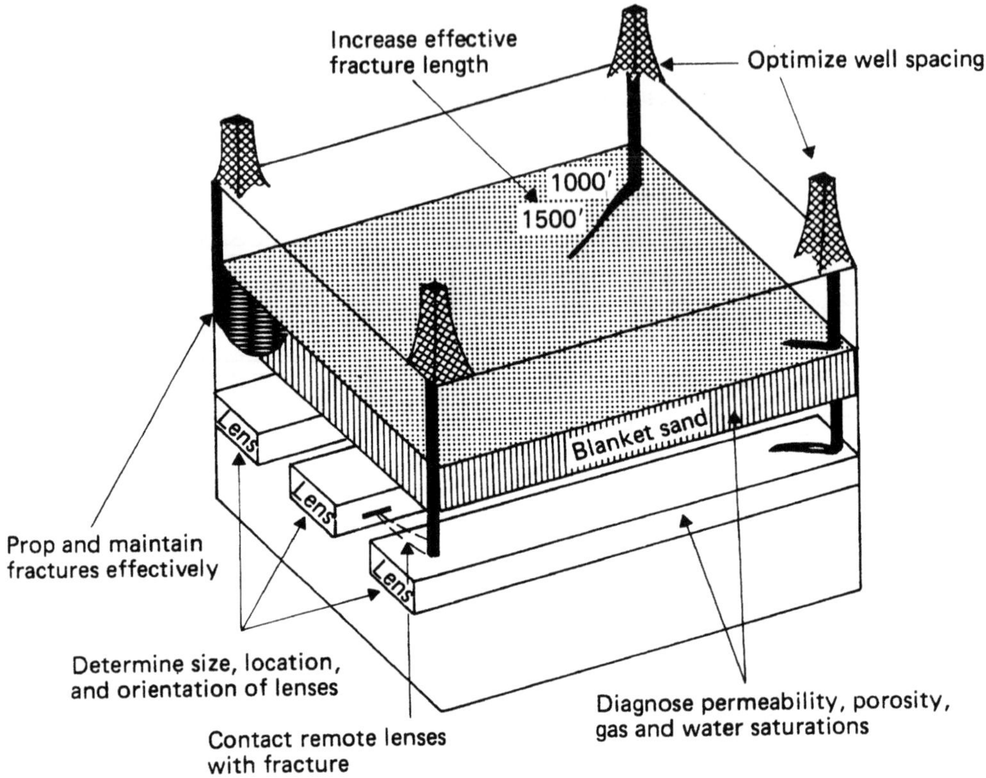

Figure 7.4. Technology advances needed to fully exploit tight gas sands.

I would like to call your attention now to a concurrent – perhaps complementary – body of research on abiogenic gas by Professor Friedmann Freund and several coworkers. Through recent ingenious and careful laboratory experiments, they have provided very strong evidence that methane and even larger H–C and H–C-N–O molecules can be catalytically formed at the surface of freshly fractured magnetic minerals, which crystallized much earlier in the presence of significant fugacities of water, carbon dioxide, and other species. Evidence of rapid mobility of dissolved carbon in this mineral was also found. Because of the huge quantities of such minerals in the Earth's crust, this could have significant implications for major abiogenic methane resources, although economic recovery of such methane is totally speculative at this time.

In any case, GRI has played a key role in the joint exploratory drilling program with the Swedish government to get a better picture of possible abiogenic gas sources. While this experiment is just getting under way, if large volumes of natural gas are found in this essentially nonsedimentary, base rock region, it would lend substantial credence to one or both of the abiogenic gas possibilities (or possibly to the ultra-deep subducted plate theory with overmatured carbon from the Norwegian trench). A strong positive find at the Siljan Crater, followed up by other appropriate research results, should begin to change some exploration strategies and, of course, could potentially affect world energy supply strategies significantly in the long run. Preliminary results are expected by the end of 1987.

Another topic of considerable importance and promise is the development of improved exploration methods. Subsurface geochemical effects associated with natural gas in the ground can lead to many possible signatures that can be detected with the right tools. Regardless of the source of natural gas, as it seeps upward, it can modify the properties of the soil or affect vegetative growth, as well as carry along other associated gases. Redox reactions can occur in minerals, sometimes with the aid of anerobic bacteria and the presence of methane.

Some very interesting work has been done using superimposed techniques; and, in a few cases, many of these have been done at the same site. One can do an instrument overfly in an airplane and look for a magnetic anomaly. The argument for this is that methane seepage coming up through the system can modify the Fe_3O_4 and Fe_2O_3 ratio in the soil. Because of near-surface oxidation of methane into carbonate ions and subsequent reaction with calcium or certain other metallic species, localized cementation occurs, filling in the soil and leading to density anomalies. This can lead to a subsequent halo effect, as porosity seals up above the reservoir and the gas travels laterally to escape.

Studies of $^{12}C/^{13}C$ ratios are quite useful indicators of inorganic versus organic sources from one area to another. Similarly, the $^3He/^4He$ ratio, and the helium level itself, are valuable tools. Uranium impurities in coal seams can relate to ^4He released, which could be associated with methane in those regions. Helium with a higher $^3He/^4He$ ratio than normal would have a good chance of being associated with abiogenic methane.

Another anomaly comes from the fact that there is a trace of uranium in many of the soils, which gives off radon. This would normally decay before it had traveled through a handful of meters of earth, given only the normal seepage times. But upsweeping gas will enhance the radon's permeation, leading to anomalously high radon concentrations. By using Terradex-type cups and various other radon detection techniques, relative hot spots can be an indicator of a possible gas reservoir.

In some soils, there is selective leaching of metallic elements associated with gas effects. Computer-processed, color band satellite measurements, coupled with ground analyses of vegetative anomalies in the plants, sometimes spot various trace element anomalies in the leaves of plants near a gas reservoir with seepage. Finally, there are electrochemical, seismic, and satellite methods using computer enhancements to detect otherwise difficult-to-spot faults, ancient meteorite damage zones, and other relevant geologic features.

Wider use of methane in the next few decades will depend heavily on technology to decrease drilling costs, particularly at greater depths and in extended reach situations. Of course, I am a proponent of aluminum drill pipe, which has several key advantages over steel pipe. Several million feet of aluminum drill pipe have been successfully used to date with new applications appearing frequently. Dr. Varnado addressed a large number of other drilling initiatives and technological opportunities for cost reduction in gas acquisition in Chapter 6, so I will refrain from elaboration on the subject.

Gas pipelines will remain the dominant mode and lowest-cost transport method for substantial quantities of gas for up to 3,000 miles, except in deep ocean or certain other hostile situations. As the size of the pipe goes up, so does its efficiency. In certain situations, transport of gas, such as LNG (liquefied natural gas), is competitive, but this constitutes only 6% of worldwide usage.

Storage of natural gas is primarily handled by pressurized underground reservoirs and, to a much lesser extent, by pipeline pressure modulation and LNG peak shaving storage systems. In the USA, developed underground storage capacity exceeds seven trillion standard cubic feet. Canada has under 400 billion cubic feet of storage; and Europe has perhaps 20 to 40 billion cubic feet of operational capacity, according to a recent study.

The historical average investment cost for such system storage capability is in the range of $3 to $4 per MMBTU capacity. Turnaround seasonal storage cost appears to average $0.10 to $0.40 per MMBTU. The maximum US gas storage daily output is roughly 40 BCF from underground storage and under 10 BCF from LNG peak shaving. There are a number of technology and safety issues involved in operating such systems.

Innumerable end-use technology opportunities are being pursued by the Gas Research Institute and various government and industry groups, but they lie beyond the scope of this chapter.

7.6. Conclusions

This chapter has described a variety of perspectives relevant to the origins and sources of gas, outlined very briefly some emerging methods of exploration and production, and touched on the relative status of pipelines and storage systems.

Although the natural gas industry is currently under economic stresses throughout much of the world, owing to depressed oil prices and a near-term supply bubble in many areas, there is a great emerging need for better understanding of our gas resource base and how to extract it more effectively in the future. Unconventional gas, including abiogenic gas, and related advances in exploration techniques and enhanced production methods will be of increased importance in the next several years, particularly when oil prices rise again.

Acknowledgment

The author would like to thank Mr. W. Gwilliam of the US Department of Energy for useful discussions and for providing the figures included herein.

References

Dienes, P. (1980), *Geochemica et Cosmochimica Acta*, **44**, pp. 943–961.
Freund, F. (1981), *Universitas*, **23**, pp. 233–239.
Freund, F. *et al.* (1983), *Bull. Mineral.* **106**, pp. 185–200.
Melton, C.E. and Giardin, A.A. (1974), *American Mineralogist*, **59**, pp. 775–782.
Wegner, A. (1915), *The Origin of Continents and Oceans*, 4th edition 1929, trans. 1966 (Methuen, London).

CHAPTER 8

Alternative Views of the Methane Age

Daniel A. Dreyfus

8.1. Methane as "The Fuel of the Future"

Considerable attention has been paid to methane or natural gas as a fuel for the future. Most of the interest, however, is the result of separate events throughout the world that are unrelated to each other or to any coherent view of a coming "methane age."

The delivery of significant quantities of gas from the Soviet Union, together with discoveries of new resources in the North Sea, have led to renewed interest in methane and competition for potential gas markets in Western Europe. Efforts to market Alaskan gas resources, deregulation of US domestic wellhead gas prices, changes in Canadian gas export policy, and development of resources in Mexico and Venezuela have made the gas trade newsworthy in the Western Hemisphere. In the Pacific, gas discoveries in Australia, Indonesia, and elsewhere have led to proposals for additional liquefied natural gas projects to expand the use of gas in Japan and to serve other industrializing Pacific economies.

This global enthusiasm for gas is further encouraged by the magnitude of the resource. Estimates of proved reserves suggest that discovered, conventional worldwide gas resources amount to over 3,000 trillion cubic feet (Tcf), or 60 times current annual production. Total conventional resources may amount to 7,500 Tcf.

The sheer quantity of the energy equivalent, dispersal of the resource, and enthusiasm to exploit it, however, will not assure that methane will become the fuel of the future. The role that gas will play within the world energy mix will depend upon the balance of its physical, technical, and economic advantages and disadvantages in relation to alternative energy forms.

Recent history, existing global infrastructure, and experience with energy trade practices and established trade relationships place the emphasis upon liquid petroleum and coal in the world energy mix. Liquid petroleum, for about 20 years, has been the dominant energy source. There are important reasons why this is so. Technically, it is a highly concentrated source of energy that can be readily tailored to the requirements of diverse energy-using devices, and it is conveniently stored and transported at ambient pressures and temperatures. In transportation uses, which place high value on energy concentration and ease of handling, petroleum products are ideal. For stationary uses, they frequently minimize the infrastructure requirements and, hence, the capital costs associated with new facilities. The extensive established world trade in oil makes the acquisition of a supply of petroleum products convenient, and reduces the investment cost for new entrants into the petroleum trade.

Between 1935 and 1965, when oil displaced coal as the dominant energy form, it was cheaply produced and sold under pricing strategies designed to expand its markets. Its growth as a fuel also coincided with the evolution of air and automotive transportation as an important energy demand sector.

Coal, which was the previous dominant fuel, is a much more difficult energy source. It is neither cheap nor easy to transport or store, and modern environmental sensitivities have added greatly to the costs associated with supplying or using it. In the vast transportation sector, it is entirely unsuited for the dominant automotive uses.

Nevertheless, when petroleum prices escalated and the security of the oil supply was called into question, the world again turned to coal. It is widely available, in its raw resource form it can be obtained at very low cost, and there is an established institutional basis for, and experience with, a world trade in coal.

For methane to expand its significance in the global energy future, it will have to gain in preference in relation to petroleum and coal. In seeking this preference, it has inherent assets and disadvantages. To some extent, advances in technology can increase the assets or mitigate the disadvantages of methane and change the potential place of gaseous fuels in the future global energy mix.

As I have said, compared with oil, methane is plentiful and generally can be found and developed at comparable prices. It is not as conveniently transported or stored, however, and current technologies for waterborne transport involve the extraordinary inconvenience and costs of liquefication. Where extensive pipeline systems already exist, incremental quantities of gas can be supplied cheaply; but in much of the world the more extensive use of gas in dispersed applications will require very high initial investments in transportation infrastructure.

Once gas has been delivered to the site, new stationary gas-using facilities may enjoy significant investment economies over coal, but the capital cost advantages of gas will depend very much upon geographic accidents of the location of the resource relative to the points of use. The economic allocation of the investment costs of the transportation system among users will strongly influence the competitive position of gas relative to other fuels.

In transportation uses, the advantages of methane as a clean-burning fuel are likely to be overwhelmed by the difficulties of on-board fuel storage.

A rational evaluation of the potential role of gas, therefore, should avoid any notion that gas is inherently a preferred or premium fuel. It is simply one of several competing energy resources. In some circumstances, it will have decided economic advantages and will be the probable choice. In others, it will be outclassed by the low cost of coal, the ready availability and physical characteristics of oil, or the overall economics of the situation.

Technologies that can expand the geographic availability of economic gas resources, reduce the cost of production, improve the economics and convenience of transportation, or enhance the end-use value of gas could change the extent to which methane becomes the preferred fuel of the future. In an extreme case, new technologies could change the fundamental role of methane.

8.2. Alternative Scenarios of the Methane Age

The dynamics by which new technologies for the discovery, production, transportation, and use of methane will be implemented will, in turn, depend upon the setting in which they must be applied – in other words, the role methane acquires in the global energy mix. The relative significance of particular technologies, as well as the time frame in which they are likely to be developed and penetrate the market, will vary according to the scenario.

In my judgment, a broad study of potential methane technologies will have to address several alternative scenarios of the "methane age," and within each scenario the study will have to consider short- and long-term time horizons. I do not believe that the more elegant and abstract methane technologies can simply be treated as longer-term technologies, which will be implemented later in the same time sequence as the more prosaic short-term technologies. Furthermore, I do not think that a true "methane age" scenario is necessarily the most interesting one to consider from a technological point of view.

As a working hypothesis, I would suggest three alternative prototypical scenarios of the methane future. For convenience I have named them:

(1) The provincial role.
(2) The world trade role.
(3) The true methane age.

8.2.1. The provincial role

The provincial role is the least revolutionary, and perhaps the most probable, scenario for methane in the global energy future. Essentially, this scenario suggests that the use of methane will expand along trends already obvious in historical experience. Methane markets will develop according to the resource opportunities and energy demand patterns of separate and dispersed geographic situations. Methane will be used primarily in traditional applications, and most international trade will be by pipeline. Waterborne trade in liquefied natural gas (LNG) will continue to be small and limited to dedicated facilities financed by contracts between producers and users to serve particular long-term contractual arrangements.

Economically, methane will generally have to be priced to compete with electricity and petroleum products at the point of use, or burnertip. Burnertip prices will have to bear the fully allocated cost of transportation, including the amortization of new investments that are required to provide transportation systems.

Because arbitrage in the well-established world oil market will tend to equalize burnertip prices for petroleum products, the wellhead price for methane will generally be a netback price that will vary according to the transportation cost. The available supply of methane for any of the disaggregated, or "provincial", markets that arise throughout the world will be limited, therefore, to the size of the resource that can be discovered and produced at the "netback wellhead price" applicable to that system.

Obviously, the ultimate scope of the global role for methane in this scenario will be significantly constrained by the outlook for world oil prices.

Intuitively, some aspects of the scenario can be sketched. European gas trade would be limited to the obvious expansion of current activity. Additional deliveries from the North Sea, the Soviet Union, and perhaps by pipeline from Africa, along with a small LNG trade, would support continued gas consumption at current or slightly increased quantities. The end-use of methane would shift over time from boilers to increased residential, commercial, feedstock, and specialized industrial process applications and peak electric power generation.

In North America, US consumption would be relatively constant, with increased pipeline imports supplementing declining domestic production. Canadian domestic demand would increase slightly. Mexico is likely to utilize most of its gas resources domestically to release oil for waterborne export. Shifts in end-use patterns would be similar to those in Europe. In the USA, greater use of gas as a fuel for electric power generation is likely, especially if acid rain considerations and aversion to the nuclear option constrain the alternatives.

Elsewhere, countries with indigenous methane resources will strive to exploit them locally, according to the opportunities offered by relative proximity to demand centers, labor force, and other resources. Gas-fired electricity generation is likely to be a significant option to avoid the cost of duplicate gas and electricity distribution systems to the ultimate users in new service areas.

Methane that is isolated from demand centers, once discovered, will have to be valued at very low prices at the wellhead. In the Pacific, LNG trade will expand only slowly,, based upon very low netback prices for otherwise unusable resources. Arctic resources will be developed very slowly.

In this scenario, most of the nearer-term supply and demand technologies currently being developed will have high priority. In general, technologies to reduce the cost of exploration and development of conventional resources and to extend the economic resource base further into frontier offshore regions, deeper horizons, less permeable formations, coal seams, etc., will be important. The objective will be to increase the quantities of methane supply that can become competitive with oil products in serving developed transportation systems.

Increasing the efficiency of methane in traditional residential, commercial, and high-valued industrial applications will also have high priority. Advanced power generation technologies, such as combined cycle power plants and fuel cells, will be increasingly sought. Technologies that can reduce the cost of constructing new pipeline transportation and distribution systems, extending existing systems, or exploiting new isolated discoveries could increase the value of the methane option.

New, nontraditional uses for gas, such as methane vehicles, would be of only specialized or geographically limited interest.

Supply technologies to provide high-cost supplements, such as synthetic gas, will be of interest only because of very long-term supply uncertainty and will receive low priority. Abiogenic or truly unconventional methane resources, however, could have a place. If the practical geographic opportunities for resource exploration were greatly expanded through these theories, the probability that indigenous gas might be discovered in places more advantageous for development and utilization would be improved.

8.2.2. The world trade role

The alternatives to the provincial scenario for the methane future are more difficult to consider because they diverge from our historical experience and depend upon less foreseeable developments in energy technology and trade.

I have called the alternative that is closest to the familiar provincial outlook the world trade role for methane. In this scenario, developments in world energy balance would encourage expansion in the international trade in methane to the extent that it would eclipse coal as the conventional alternative to petroleum. This scenario would depend upon the extension of pipeline transportation systems wherever continental resources exist and the development of an extensive waterborne trade, presumably in LNG. A waterborne trade of sufficient capacity and with institutional brokerage arrangements would be necessary for a shipper or buyer to be able to enter the market without negotiating and financing a complete dedicated supply system.

The end-uses of methane would still be largely traditional ones, but the potential markets would be geographically expanded well beyond today's historical gas-using communities.

Intuitively, this scenario implies:

(a) The extension of pipelines in North America to exploit the Arctic resource and the Mexican fields.
(b) An LNG trade, both inter-American and trans-Pacific, that could sustain greater long-term methane use in the USA.
(c) The development of gas distribution systems throughout metropolitan areas of Central and South America.

In Europe, similarly, extensions of pipelines and expanded LNG trade could increase the significance of gas in relation to oil and coal in the energy mix. Throughout the rest of the world, isolated gas resources would be exploited for pipeline or waterborne trade to high-value markets, rather than consumed on site for whatever uses the local opportunities permit. Methane would have a larger share of the energy mix in the industrialization of Pacific Rim and Indian subcontinent economies.

The probable economies of methane in this scenario may be derived from the circumstances. The scenario can only evolve if the burnertip value of gas is sufficiently great to encourage the tremendous investment required for the global transportation infrastructure. The circumstances, therefore, must be that the world price of petroleum, including the imputed costs of political insecurity and future uncertainty, become very high, and the costs of expanded coal use also increase. The global coal resource, however, is extensive and found in widely divergent political circumstances, and it can be produced cheaply. The other condition that would be necessary to create an economic window for gas, therefore, must be environmental constraints on coal use, either through direct restrictions or economic penalties derived from pollution controls.

Methane, therefore, would be seen as replacing oil in its less valuable industrial applications and displacing coal in some industrial and electricity generation applications through an economic advantage arising from environmental constraints. Methane would also be chosen to reduce end-use electricity demand in residential, commercial, and some industrial applications, either for direct policy reasons or as the result of an economic advantage over coal-fired generation.

The paramount technologies to facilitate this world trade role for methane are likely to be:

(a) Electric power generation technologies, including advanced cogeneration concepts.

(b) End-use technologies to expand the usefulness of gas in industrial applications, space cooling, and feedstocks.
(c) Transportation technologies, including LNG system improvements.

Large-scale storage technologies could be important to balance the large investment in LNG and pipeline supply capacity with seasonal fluctuations in demand at an optimum economic level. The notion of some inherent constraint on the petroleum trade also enhances the potential for methane-fueled transportation options in this scenario.

The implied higher value for gas in this scenario, increased global consumption, and, hence, the greater exploitation of the resource base will increase the significance of exploration and production technologies. The potential need for long-range supplements will also increase interest in synthetics.

Truly unconventional or exotic methane resources would remain interesting, but only if they could be exploited in competition with relatively high-cost, but available, coal and oil, or in the very long term (post-2010).

8.2.3. The true methane age

My third and final scenario is the achievement of a true "Methane Age". In this scenario, methane would replace liquid petroleum as the dominant global energy source. This is the least familiar and most difficult scenario to consider, and significant divergence from current trends would be necessary for it to come about within the next 50 years.

The most logical development of this scenario would be for gas substantially to replace coal and oil in stationary energy applications worldwide. Gas would have to become the preferred primary fuel for baseload electric power generation, presumably as the result of constraints on the use of coal and nuclear energy, rather than solely through advances in gas technologies. A convenient global transportation system would have to be available, and long-term economic supplies of natural gas and supplemental gaseous fuels would have to be assured to underwrite the investment involved.

Such a scenario could result from technological advances that greatly increase the electrification of end-use energy applications. Extreme increases in electricity demand would, in turn, multiply the requirement for primary fuel. If coal and nuclear generation were constrained and liquid petroleum prices high, methane could become a fuel of choice for electric power generation. The technical advantages of methane in decentralized generating facilities could be optimized to supplement large, centralized coal and nuclear baseload generating plants. Power generation in relatively small combined cycle power plants near demand centers, along with cogeneration and fuel cells, would convert methane to electricity at the point of use. These facilities would be supplied with gas by pipeline, but final distribution of energy to small users would be in the form of electricity.

Cogeneration can economically provide baseload power, and gas turbines are an economical form of peaking and backup capacity.

A true methane age would be dependent upon significant advances in technology. In order for methane to displace other fuels in a major portion of end-use energy service needs, even stationary needs, efficient and flexible conversion to electricity will be necessary over the widest range of services. Methane will have to capture a significant share of the primary fuel consumption for electric power generation, including baseload service.

Technologies for the storage of methane throughout the transportation system and on site at the point of consumption will need to advance. Advanced technologies for direct conversion to electricity and for feedstock applications of methane will also be required to expand the potential applications.

In general, the widest variety of possible energy demands will have to be addressed so that the immense investment necessary to establish the global supply system can be justified and amortized. It is difficult, even in a methane age scenario, to envision that gas will become the dominant transportation fuel so long as liquid petroleum products are readily available. Conversion of methane to liquids, however, will become interesting for the specialized transportation and storage of methane as an energy source and, eventually, for on-board transportation fuels once extensive supply networks are available. It is questionable, in fact, whether a true methane age will be achievable without a significant share of transportation markets. The nature of the transportation demand may effect a necessary balance with other demand variations that influence the transport and storage of methane.

On the supply side, similarly, the widest possible quantity and geographic dispersal of methane sources must be developed. Technologies for exploration and development of all frontier sources, including deep water offshore, deep horizons on and offshore, and difficult climatological regimes, must be conquered. The conventional resource will have to be extended and supplemented by development of unconventional resource bases to their ultimate potential so that the high demand levels can be sustained with considerable confidence in very long-term supply security.

The potential for abiogenic methane, development of hydrate sources, and, ultimately, the technologies for replacement of the methane economy with hydrogen will become significant.

8.3. The Relevance of an IIASA Study

Clearly, the technologies of interest, and especially the priority and dynamics of particular technology implementations, would differ among these three prototypical scenarios. It would be futile to consider specific methane technologies, the technical and economic potential for their utilization, and the dynamics of their potential implementation without consideration of the general role that methane is expected to play as the technologies evolve.

For example, in the early 1970s in the USA, natural gas was perceived to be in short supply. The demand for gas was increasing largely because its regulated price became artificially low relative to the prices of alternative fuels during the petroleum crisis era. Several major initiatives involving LNG facilities and synthetic gas plants were implemented, predicated upon a consensus scenario of a greatly expanded future role for gas. It was anticipated that gas would continue to enjoy a competitive price advantage in comparison to oil, along with policy incentives for its use.

The analytical basis for the natural gas outlook was incoherent. It included the following somewhat inconsistent assumptions:

(1) Domestic US gas production would continue to diminish despite rising wellhead prices.
(2) High-priced gas supplements could be sold in competition with petroleum products and electricity at the point of use.
(3) Gas, as a preferred resource, should be constrained by policy to the residential and commercial markets, although those markets required only 8 Tcf per year of the 22 Tcf being consumed annually in the mid-1970s.

As the scenario played out, higher gas prices brought forth greater quantities of conventional supplies. High prices, policy constraints, and a history of sporadic supply curtailments stifled industrial demand for gas. Finally, excess domestic supplies and competitive Canadian imports resulted in depressed prices, together with surplus supplies and stagnant demand.

The LNG, Alaskan, and synthetic supplements are no longer of immediate interest. Planned and budgeted projects that were intended to be in service in the 1980s have become abstract notions for the next century.

The dynamics of new methane technologies will be just as dependent upon circumstances. The potential technologies are only important in the sense that they shape the physical and economic limitations for the scenarios. A particular technology may be necessary for a scenario to come about, but it will not insure the outcome. The economic setting, alternative energy sources, and demand patterns will all play a role. When various emerging or potential methane technologies are addressed without the discipline of environmental settings, the study is likely to become internally inconsistent and abstract. Abiogenic methane or advanced electricity conversion technologies, for example, are likely to have very different significance or potential for implementation, depending upon the sophistication of the methane transportation and storage system that evolves. Furthermore, the role of methane relative to other fuels will place constraints upon the technical and economic criteria for the successful implementation of new technologies.

IIASA's study can make its greatest contribution to deliberations about the outlook for methane technologies by:

(1) Describing a range of scenarios for the role of methane in the world energy mix, including careful consideration of the implications for other energy sources and for methane markets that are inherent in each scenario.
(2) Within a consistent set of economic and technical assumptions established for each scenario, tracing the practical dynamics of introducing potential new methane technologies and the technological and economic requirements for these technologies to support the role of methane within the scenario.
(3) Evaluating various emerging and potential methane technologies with respect to their probable availability as required for the scenarios to come about.

CHAPTER 9

Combined Cycle Systems: Technology and Implications

Thomas H. Lee

9.1. The Combined Cycle System Today

In a gas turbine combined cycle system (GTCC), the exhaust from a gas turbine is fed into a residual heat boiler that generates steam for a bottoming steam turbine cycle. If natural gas is used to fuel the gas turbine, the overall efficiency for commercial GTCC systems is about 47%. The capital cost of such a system is about $400/kW. (Unlike estimates of nuclear plant costs, this pricetag is quite accurate.) Yet, with these attractive numbers, combined cycle systems have not been considered a serious option in planning additions to utilities' future power generation. The reasons are many:

(1) The Fuel Use Act, which forbids the use of natural gas for electricity generation in the USA.
(2) The lack of confidence in reliable long-term gas supply.
(3) The lack of confidence in the performance of GTCC.

The last point is a rather interesting one. For many years, heavy-duty gas turbines were used mostly for handling peak loads. Their availability or reliability was never as important as that of baseload units. In addition, the outage time was usually rather short, i.e., if something went wrong, it did not take very long to fix it. However, when combined cycles using these gas turbines were put in place for baseload application, reliability became an important issue. Also, GTCC control systems did not receive as much attention as they should have. As a result, the overall performance of combined cycles in the late 1970s was indeed rather unsatisfactory.

But the problems that caused poor reliability were clearly solvable. Another type of gas turbine – the jet engine – was capable of running 50,000–100,000 hours without repair. The technology to build reliable control systems was also well known. If sufficient effort had been applied to improve the reliability of combined cycle systems, the probability of success would have been very high. But because there was no market, manufacturers had little incentive to tackle the problem.

Two things happened that changed the situation. First, the Electric Power Research Institute (EPRI), in cooperation with Southern California Edison and Texaco, proposed a consortium to develop a $300 million integrated coal gasification combined cycle system – the Coolwater Project. General Electric, the leading gas turbine supplier, joined the consortium and took responsibility for the combined cycle portion of the system. For the first time, the reliability issue of combined cycle systems was addressed in detail.

Second, Japan, in an effort to diversify energy sources, ordered several GWs of combined cycle systems using liquefied natural gas. Suddenly, combined cycle systems came into the spotlight.

In the author's laboratory at MIT, we decided to take a careful look at the potential of combined cycle systems in future power generation planning, a subject which had received very little attention so far. We used for this analysis the newly released Electric Generation Expansion Analysis System (EGEAS), developed by MIT and Stone and Webster Engineering Corporation for EPRI (Caramanis *et al.*, 1982). EGEAS involves planning analyses using one of three optimization structures: linear programming, dynamic programming, and generalized Bender's decomposition.

In order to evaluate the GTCC technology within a range of utility environments, Tabors and Flagg (1986) used the EPRI-developed Regional Utility Systems. The Regional Utility Systems provide scaled-down models of existing capacity and load duration curves as well as fuel price regimes for each of six continental US regions. *Figure 9.1* and *Table 9.1* summarize the characteristics of each of the six regions. Results of analyses carried out for the model regions may be "expanded" to reflect actual national or regional impacts of multiplying by the scale factor indicated in *Table 9.1*. The study was carried out for a 15-year planning horizon, beginning January 1, 1990, and ending December 31, 2004.

With a set of reasonable assumptions, the study concluded that natural gas-fired combined cycle systems offer an important alternative for capacity planning. In all regions they contributed to the optimal capacity mix. In three of the six regions, they provided the majority or all of the optimal mix. The results of the regret analysis show clearly the cost implications of not allowing for the natural gas-fired combined cycle option. Further, these studies point clearly to the minimal downside risk associated with significant capacity investment in the GTCC technology.

The study concluded that:

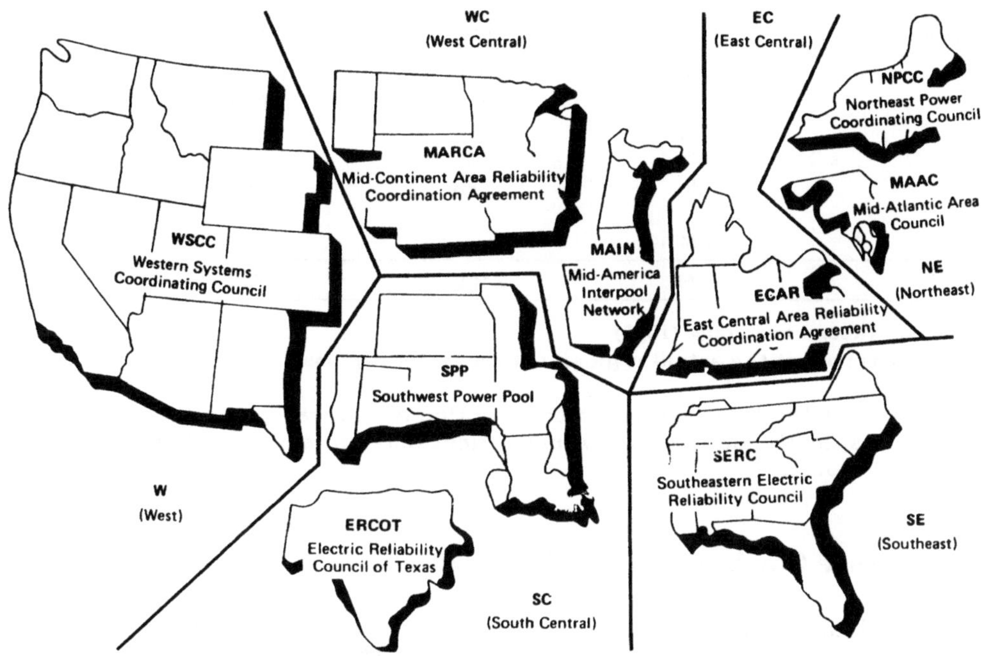

Figure 9.1. EPRI regional systems: six regions based on National Electric Reliability Council Regions (contiguous USA).

Table 9.1. Regional utility characteristics (%)

Region	*Hydro*	*Nuclear*	*Coal*	*Oil and Gas*	*CT*	*Growth rate*	*Res margin*	*Scale factor*
NE	10	21	21	36	12	2.1	36	5.74
SE	8	14	46	12	10	3.7	29	7.90
EC	0	12	76	4	8	3.5	33	5.79
SC	1	10	41	46	2	4.2	20	6.70
WC	3	22	59	8	8	3.2	19	4.34
W	33	13	29	19	6	3.5	35	7.04

Source: Tabors and Flagg (1986).

Current governmental policy does not allow for natural gas to be used in new power generation. This decision was based on what now appear to be faulty assumptions concerning gas availability and clearly on faulty assumptions concerning fuel economics. Further, *Table 9.2* indicated the projected total increase in natural gas consumption (scaled to the U.S.) required to supply the optimal capacity expansion plans defined in this analysis. It is striking to note that relative to the DOE/EIA projections which assume no new gas capacity, the projections presented in this paper reflect only an increase from 4×10^{12} to 7.8×10^{12} SCF or less than a doubling in utility gas consumption. In

Table 9.2. Total incremental gas consumption by region (1×10^9 SCF).[a]

	Year			
Region[b]	*1990*	*1994*	*1999*	*2004*
NE	419	1,039	1,688	2,411
SE	79	711	1,904	3,350
EC	0	226	359	643
SC	375	281	342	643
WC	26	126	395	707
W				
S	2004			
Total	899	2450	4700	7780

[a]Source: Tabors and Flagg (1986).
[b]The values reflect the actual regions, not the model regions.

terms of the total projected consumption of natural gas in the United States, in all sectors, this would reflect only an increase of 20% over 1995 levels.

9.2. Future Combined Cycle Systems

The development of gas turbines for electricity generation was heavily influenced by jet engine technology. The US government, over several decades, invested hundreds of millions of dollars annually to improve the performance of jet engines. Firing temperatures have increased (*Figure 9.2*), high-temperature alloys were developed (*Figure 9.3*), and cooling systems were improved. All these advances were exploited in the development of heavy-duty gas turbines.

For aviation applications, size and weight are significant constraints. This is not true for ground applications. Thus, an optimum design for an aviation application may not be the optimum for electricity generation. Moreover, an optimum design for a free-standing gas turbine may not be the optimum for combined cycle systems. These are surprisingly recent insights.

In the following, we review briefly the status of combined cycle R&D (Magnusson, 1982).

The simple cycle industrial gas turbines in use today operate at firing temperatures of 1,144–1,422° K (1,600–2,100° F) and cycle pressure ratios (CPRs) of 8 to 14. These machines achieve simple cycle thermal efficiencies of up to 32.5%. Gas turbines in combined cycle arrangements with a steam turbine can achieve thermal efficiencies in the neighborhood of 45%. The next generation of industrial gas turbines is now beginning to be sold commercially.

According to the NASA-sponsored report by Robson (1981), the most advanced, state-of-the-art industrial gas turbine offered today is the FT50/GT200, developed by United Technologies Corporation and Stallaval of

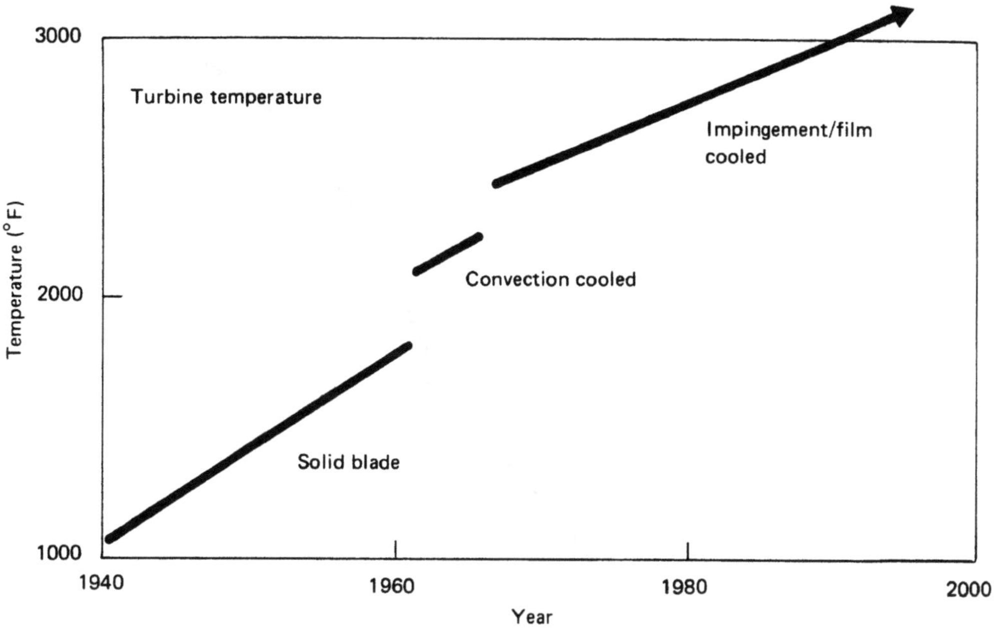

Figure 9.2. Turbine technology trends.

Sweden. The air film-cooled turbine incorporates many aircraft turbine advances into an industrial turbine operating at 1,422° K (2,100° F) with a CPR of 16 and an achieved thermal efficiency of 34%. Other manufacturers are also offering updated versions of their industrial gas turbines. General Electric's new version of the Frame 7 operates at 1,422° K (2,100° F) and a CPR of 11.5. Westinghouse's W501D operates at 1,403° K (2,060° F) and a CPR of 14.2 (Robson, 1981). The commercially available gas turbines all require clean fuels, such as natural gas, to achieve the thermal efficiencies cited above.

Present gas turbine combined cycles are limited by low turbine exit temperatures (TET), and the exhaust cycle is driven by steam at low temperatures and pressures. Most steam cycles in combined cycle arrangements operate at pressures of 100 Atm. (1,465 psia) and temperatures of 700° to 756° K (800 to 900° F). This is considerably less advanced than current fossil fuel-fired steam plants that operate at pressures of about 164 Atm. (2,415 psia) and temperatures of 811° K (1,000° F). With new gas turbines operating at higher inlet temperatures, resulting in higher exhaust temperatures, combined cycle steam conditions can be improved.

The logical starting point in identifying the more promising concepts for advanced power cycles is to locate the sources of inefficiency in existing technologies. A power plant may be regarded, thermodynamically, as a system that separates energy from entropy. The end product, work, is energy free from entropy. The initial entropy in the energy source plus that generated in the process must be rejected, by necessity involving

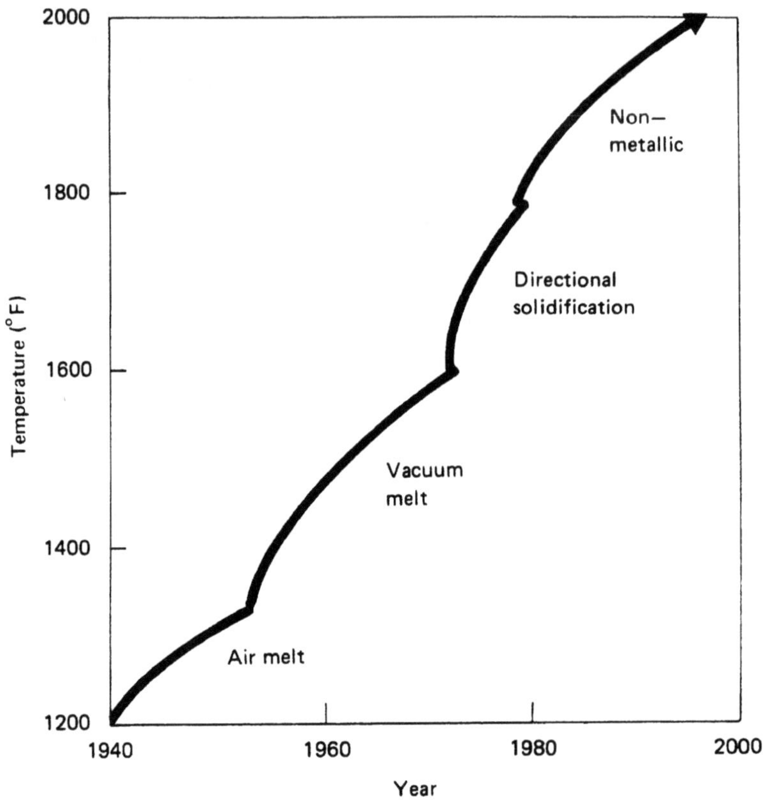

Figure 9.3. Progress in turbine blade and disk materials.

energy wastage. With this in mind, an analysis of the source of loss in a power plant must be based upon the second law of thermodynamics.

To illustrate this philosophy, *Table 9.3* was prepared, showing the energy (first law) and availability (second law) losses and their distribution in a typical modern steam power plant. A look at the energy loss figures

Table 9.3. Typical energy and availability losses (%) in a modern steam power plant.

Process	*Energy loss*	*Availability loss*
Combustion	1	30
Heat transfer	5	23
Exhaust flue	10	2
Condenser	49	2
Mechanical equipment	5	5
Useful work	35	38
Total	100	100

may lead one to believe that there is very little room for progress. The major loss, condenser heat rejection, may seem unavoidable. Examination of the availability loss column, however, reveals the real sources of inefficiency and suggests the avenues for improvement.

9.3. Rationale for the Proposed System

Since the major sources of loss are the irreversibility of the combustion process and the heat transfer from hot combustion gases to boiler temperatures, improving those processes should yield the greatest increases in efficiency.

Combustion irreversibility could be reduced if the process were carried out gradually and at high temperature. When temperatures are sufficiently high for equilibrium between reactants and products, the process becomes close to reversible. The limitations are (1) the unavailability of semipermeable membranes to introduce a component into the mixture at its partial pressure and (2) the necessity for maintaining sufficient disequilibrium to achieve a finite, acceptable reaction rate. Both those losses can be relatively small. The reaction may be kept close to equilibrium only by a simultaneous process of reactant mixing and energy extraction to maintain the required temperature.

The problem of availability loss in heat transfer can be overcome by extracting work directly from the high-temperature gases rather than degrading the energy to a lower temperature. This may be done in a device such as gas turbine.

9.4. The Interstage Combustion Turbine Cycle

One method of meeting the above-mentioned requirements is to build a combined cycle consisting of an interstage combustion isothermal gas turbine and a bottoming steam cycle.

To illustrate the thermodynamic benefits of this approach, *Figure 9.4* compares four cycles running between 300° K and 1600° K, except for the steam cycle whose peak temperature is 920° K. Values of theoretical and typical efficiencies are noted. The isothermal gas turbine with a well-matched bottoming steam cycle is shown to approximate the Carnot Cycle.

The efficiency of the proposed cycle increases with turbine temperature. However, this necessitates cooling the gas turbine blades, which introduces thermodynamic losses. Achieving this cycle also requires gas turbine pressure ratios well above the capability of current technology.

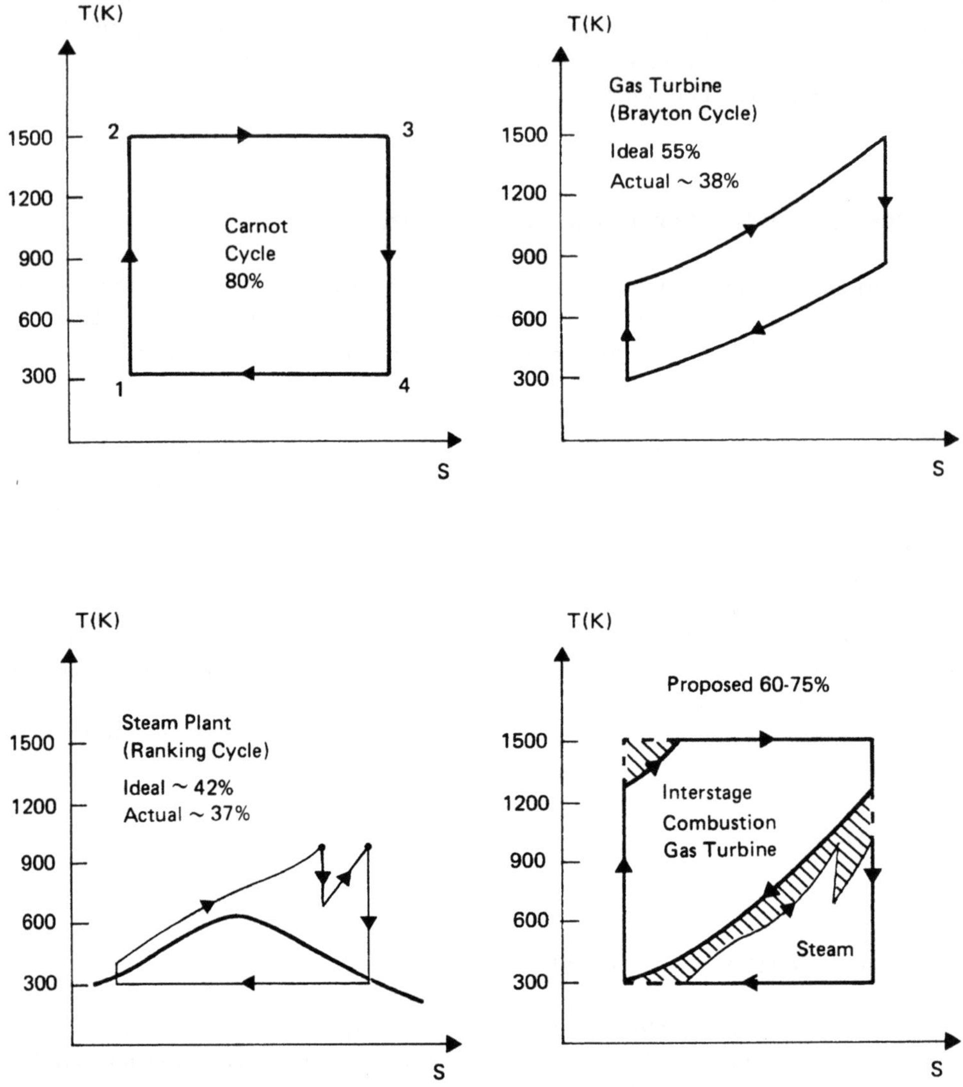

Figure 9.4. Four cycles consisting of an interstage combustion isothermal gas turbine and bottoming steam.

9.5. Reheat Gas Turbine Cycles: The Japanese Moonlight Project

A national energy savings project started by the Japanese government in 1978 includes a full-scale research and development effort aimed at building an advanced reheat gas turbine combined cycle, burning LNG, and achieving an efficiency of 55% (LHV). The cycle will operate at very high pressures, 55 Atm., but the firing temperatures are in line with current practice:

1,573° K (2,372° F) for the turbine exhaust temperature (TET), and 1,473° K (2,192° F) for the reheat combustor. The test on the first prototype began in 1984.

The high pressure ratio used in the Moonlight Project design is prompted by the fact that reheat gas turbines need a much higher CPR to achieve a TET that is compatible with the exhaust steam cycle. Bottoming steam cycle conditions will be 164 Atm. and 839/839° K (2,415 psia and 1,050/1,050° F) driven by a gas turbine exhaust at 893° K (1,147° F) (Hori and Takeya, 1981; see also Magnusson, 1982).

The proposed cycle is a dual-shaft system with a high-pressure shaft and a low-pressure shaft arranged in tandem. Mounted on the high-pressure shaft and rotating at free speed is the high-pressure compressor and the high-pressure turbine separated by the primary combustor. The low-pressure system, mounted on its separate shaft rotating at a fixed speed, includes the intermediate-pressure turbine, low-pressure turbine, and the generator. The reheat combustor is located in between the intermediate- and the low-pressure turbine. *Figure 9.5* (Hori and Takeya, 1981) shows a schematic of the reheat gas turbine combined cycle plant. Raising the initial firing temperature to 1,773° K (2,732 °F) to further increase efficiencies is contemplated.

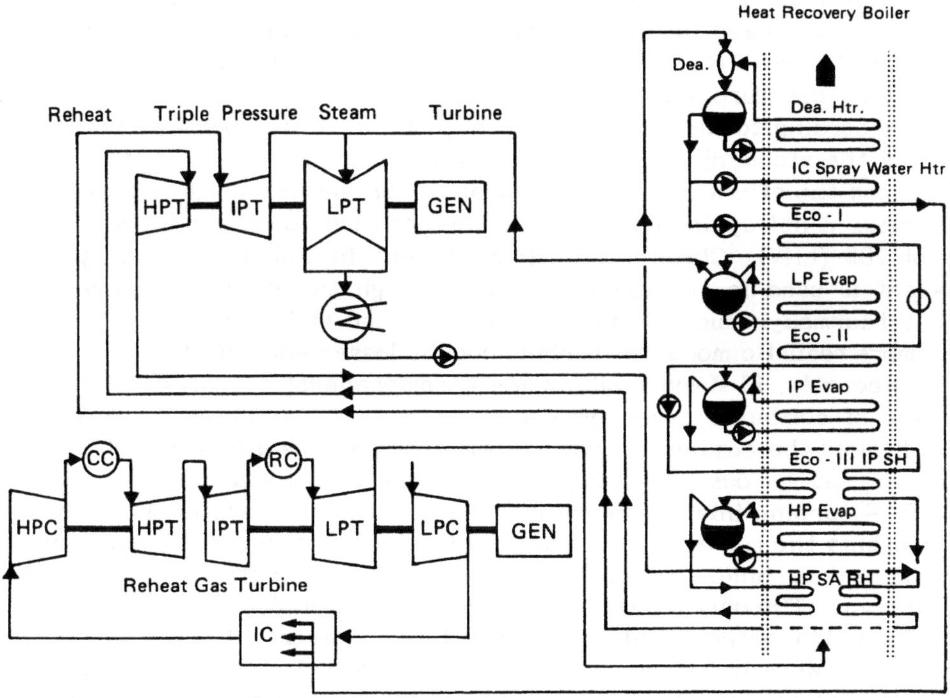

Figure 9.5. **Schematic of the Japanese Moonlight Project reheat gas turbine combined cycle power plant (Hori and Takeya, 1981).**

9.6. The Steam-Cooled Reheat Gas Turbine Combined Cycle

This cycle has been proposed by Rice (1981a). Based on the new developments of the E3 (Gardner and Grey, 1980) and the HTTT (Horner and Caruvana, 1982) programs, this cycle combines a gas turbine reheat combustor, similar to the one developed by the Japanese, with an elaborate steam cooling scheme. The cycle proposed by Rice would operate at a CPR of 40, a TET of 1,700° K (2,600° F), and a gas turbine reheat temperature of 1,478–1,589° K (2,200–2,400° F) combined with a 164 Atm. 867/839/714° K (2,415 psia 1,100/1,050/825° F) double reheat steam cycle.

A steam-cooled version of the Pratt and Whitney E3 program engine would be used as the cycle gas generation unit. The gas generator would exit into an annulate steam-cooled reheat combustor and beyond into the steam-cooled power turbine. *Figure 9.6* shows a schematic of the cycle concept.

The cycle cooling system is optimized to enhance the steam cycle end of the combined cycle. The steam flow is split 50/50 before the first steam cycle reheat. Half goes to cool the reheat combustor, and half is used to cool part of the power turbine. According to Rice, this heats the steam to the required reheat temperature of 1,050° F. After the second steam expansion in the intermediate-pressure turbine, the steam flow is again split. Some 42% of the steam goes to cool the rest of the power turbine and is thereafter mixed in with the mainline steam in the crossover line/mixer at "0" in *Figure 9.6*. This would raise the steam temperature to the second reheat temperature of 850° F.

According to Rice, it is the cooling of the reheat part of the turbine, raising the steam cycle reheats to their full temperature, that so greatly enhances the steam cooling concept and improves the combined cycle efficiencies. A net cycle efficiency for the optimized cooling case is estimated at 58.78% (LHV). Clearly, a cooling concept in which the energy of the coolant is used in the exhaust cycle at high temperatures improves the cycle thermodynamics. However, the estimated efficiencies would be lower if a more realistic model for turbine cooling losses were used.

Recently, El-Masri (1985) made a comprehensive analysis of the efficiency of combined cycle systems. His results are shown in *Figure 9.7*, which plots the combined cycle system efficiency as a function of pressure ratio (r_p) at various peak temperatures (measured by Θ, which is the ratio of TIT – turbine inlet temperature – to ambient temperature). Each point represents the maximum efficiency configuration of single-stage compressor and reheat turbine. The optimum number of turbine stages is indicated at each point. The only exceptions are the points for $\Theta = 4$ and $\Theta = 5$ with $r_p = 128$, which optimize with intercooled compressors. It is worth nothing that, if we were to restrict the maximum number of reheat stages to three, say, the cycle with $\Theta = 24$, $r_p = 64$ would also optimize with compressor intercooling.

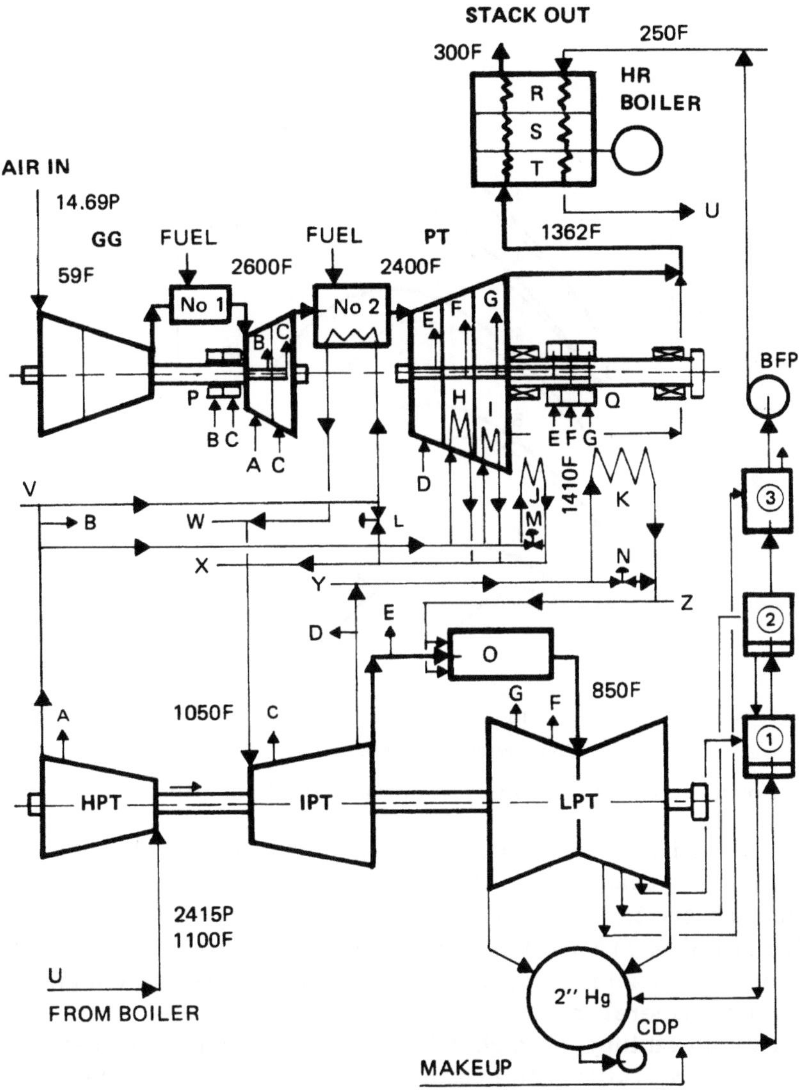

NOMENCLATURE

COOLING STEAM - OPEN CYCLE

A. 1st STG GG VANES
B. 1st STG GG BLADES
C. 2nd STG GG VANES & 2nd STG GG BLADES
D. 1st STG PT VANES
E. 1st STG PT BLADES
F. 2nd STG PT BLADES
G. 3rd STG PT BLADES

COOLING STEAM - CLOSED CYCLE

H. 2nd STG PT VANES
I. 3rd STG PT VANES
J. PT SHELL AND STRUTS
K. PT EXHAUST HOOD AND DUCT

REGULATORS, SEALS & MIXER

L. RM COMBUSTOR PRV
M. PT VANES, SHELLS & STRUT PRV
N. PT EXHAUST HOOD & DUCT PRV
O. CROSSOVER MIXER TO LPT
P. GG PACKING
Q. PT PACKING

HEAT RECOVERY BOILER

R. ECONOMIZER
S. EVAPORATOR
T. SUPERHEATER

STEAM HEADERS

U. HP 2415 P
V. COLD RM 604 P
W. RH 545 P FROM RH COMB
X. RH 545 P FROM PT
Y. COOLING STEAM 140 P to PT
Z. HOT RETURN 123 P to MIXER

Figure 9.6. **Schematic of the steam-cooled reheat gas turbine cycle proposed by Rice (1981a).**

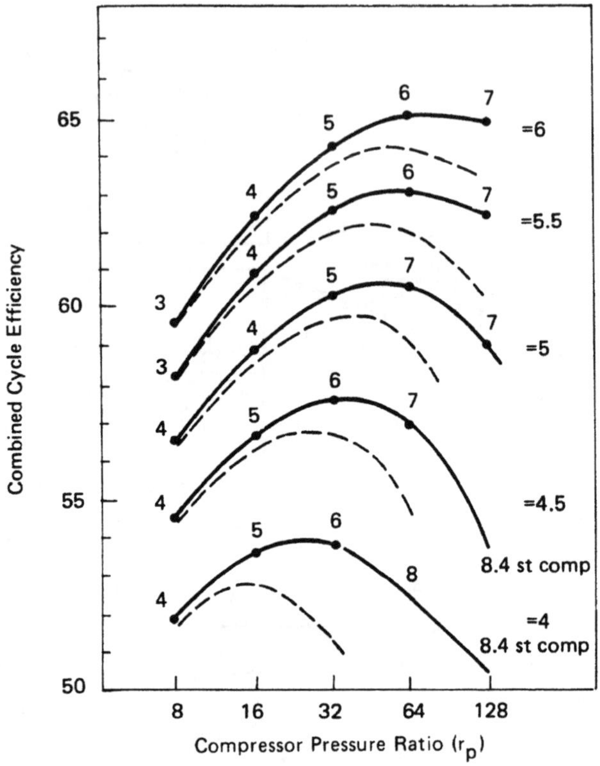

Figure 9.7. **Combined cycle efficiency at optimum configuration.**

Owing to design constraints, it would be difficult to construct a machine with more than three stages (two reheats). To investigate the effect of this restriction, a configuration of single compressor with three turbines is plotted in *Figure 9.7*. These dashed curves fall below the solid curve for optimum efficiency by an amount that increases at higher pressures and lower temperatures.

When the possibility of high-efficiency combined cycle systems was included in the EGEAS study, the results were indeed remarkable. The most dramatic differences occurred in the Northeast and the Southeast regions. *Figures 9.8* (*a* and *b*) show the expansion paths for these two regions, in which only natural gas-fired combined cycle systems are added to capacity over the length of the planning horizon. *Table 9.4* presents the 15-year time stream of capacity additions for the Northeast. The model forecasts a total of 12,250 MW of combined cycles for the scaled regional utility. This expands to over 70,000 MW for the actual geographic region extending west through Indiana and south through Kentucky. Of this total, 70% is required to replace scheduled retirements.

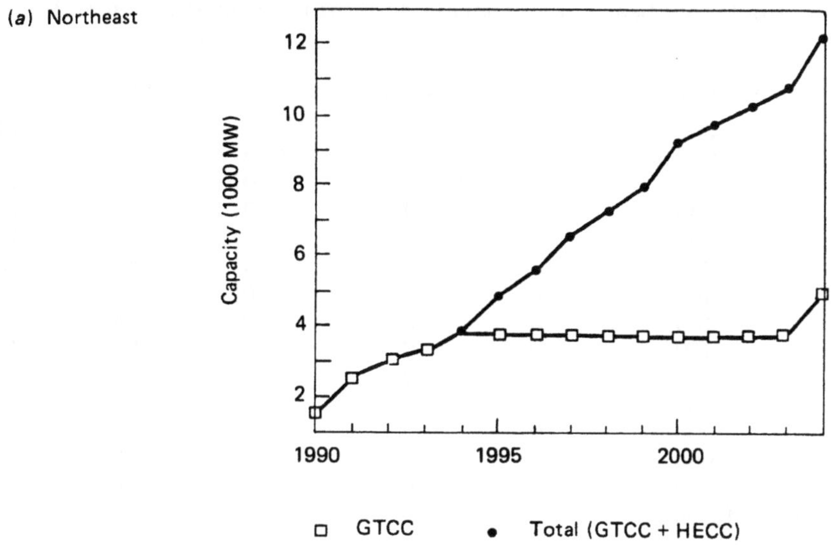

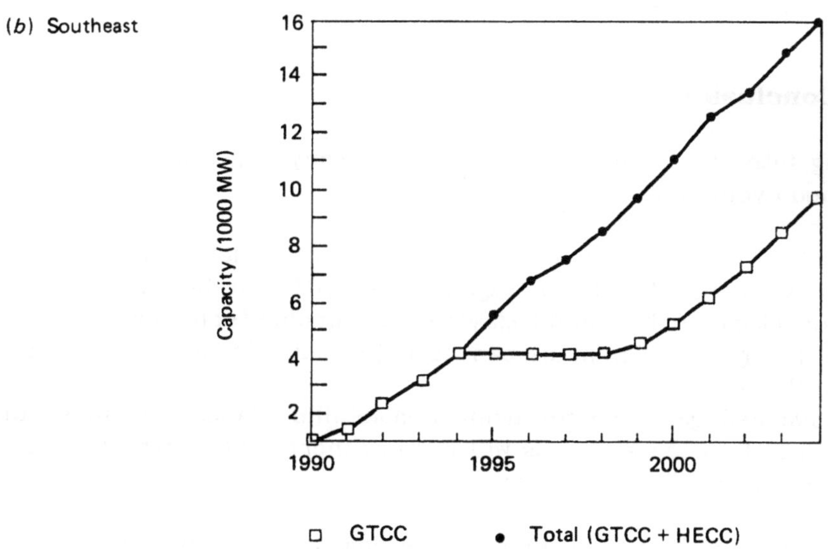

Figure 9.8. Projected capacity of high-efficiency combined cycle systems in two regions: (*a*) Northeast, and (*b*) Southeast.

For the other four regions, the combined cycle systems share predicted capacity additions with Atmosphoric Fluidized Bed Combustion (AFBC) in the West Central, South Central, and East Central regions. Only in the West was the combined cycle system not as attractive.

Table 9.4. Northeast base case: annual additions to generating capacity.[a]

Year	*Retirements*	*NUC*	*AFBC*	*GTCC*	*HECC*	*Total capacity*
1990	800	0	0	1,500	0	20,500
1991	800	0	0	1,000	0	20,700
1992	100	0	0	500	0	21,100
1993	600	0	0	250	0	20,750
1994	600	0	0	500	0	20,650
1995	600	0	0	0	1,000	21,050
1996	400	0	0	0	750	21,400
1997	400	0	0	0	1,000	22,000
1998	400	0	0	0	750	22,350
1999	300	0	0	0	750	22,800
2000	1,100	0	0	0	1,250	22,950
2001	300	0	0	0	500	23,150
2002	900	0	0	0	500	22,750
2003	200	0	0	0	500	23,050
2004	1,200	0	0	1,250	250	23,350
Total	8,700	0	0	5,000	7,250	23,350

[a]Tabors and Flagg (1986).

9.7. Conclusion

Looking into the future, we can imagine many important advances of the combined cycle system technology:

(1) Reheat gas turbines are already under active development.
(2) Conversion efficiencies higher than 50% can be achieved with the reheat gas turbine and today's steam turbine technology.
(3) Still higher efficiencies can be expected with more advanced cooling methods.
(4) Advanced gas turbines using nonmetallic blades are under development. If successful, this will further increase the overall efficiency of the combined cycle systems

Whether methane will become the dominant energy source depends on two factors: its availability and progress in related technologies, one of which – combined cycle systems – shows great promise.

References

Caramanis, M.C., Schweppe, F.C. and Tabors, R.D. (1982), *Electric Generation Expansion Analysis System*, Vol. 1, EPRI Report EL–2561 (Electric Power Research Institute, Palo Alto, CA).

El-Masri, M.A. (1985), On thermodynamics of gas turbine cycles, *ASME Transactions* **107**, October.

Gardner, W.B. and Gray, D.E. (1980), *The Energy Efficient Engine (E3) – Advancing the State of the Art*, ASME Paper No. 80-GT-142.

Hori, A. and Takeya, K. (1981), *Outline of Plan for Advanced Research Gas Turbine*, ASME Paper No. 81-GT-28.

Horner, M.W. and Caruvana, A. (1982), *High-Temperature Turbine Technology Readiness*, ASME Paper No. 82-GT-213.

Magnusson, J.H. (1982), *A Gas Turbine Combined Cycle Approaching Minimal Irreversibility*, M.S. thesis (MIT, Department of Mechanical Engineering, Cambridge, MA).

Rice, I.G. (1981a), *The Reheat Gas Turbine with Steam-Blade Cooling – A Means of Increasing Reheat Pressure, Output and Combined Cycle Efficiency*, ASME Paper No. 81-GT-30.

Rice, I.G. (1981b), The reheat gas turbine combined cycle, *Turbomachinery International*, July–Aug.

Rice, I.G. (1982a), The reheat gas turbine combined cycle, *Mechanical Engineering* **104**(4), April: 46–57.

Rice, I.G. (1982b), *The Reheat Power Turbine with Steam Cooling – A Means of Increasing Reheat Temperature, Exhaust Temperature and Combined Cycle Efficiency*, ASME Paper No. 82-GT-134.

Rice, I.G. and Jenkins, P.E. (1981), *Comparison of the HTTT Reheat Gas Turbine Combined Cycle with HTTT Nonreheat Gas Turbine Combined Cycle Efficiency*, ASME Paper No. 81-GT-69

Robson, F.L. (1981), Applications study of advanced power generation systems utilizing coal derived fuels. Volume II: Technical Report. *Final Report*, United Technologies Research Center, NASA-CR-161692, N81-21539, March.

Tabors, R.D. and Flagg, D.P. (1986), Natural gas fired combined cycle generators: Dominant solutions in capacity planning, *IEEE Transactions on Power Systems*, Volume PWRS-1, (2), May: 122–127.

CHAPTER 10

The Engineer's Contribution to the Methane Age

Tom Schur and Jack Metz

10.1. Sulzer Brothers

Sulzer Brothers was founded in Winterthur, Switzerland, in 1834 as a metal foundry, but soon turned to making steam engines. In fact, the Sulzers went to England to recruit one of the early steam engine pioneers, Charles Brown by name, and with his help, built their first steam engine in 1843. (Charles Brown's son, incidentally, later became the cofounder of Brown Boveri.) By the end of the nineteenth century, Sulzer had attained quite a high reputation for its engineering products. It was not surprising that a certain well-to-do family from Germany wanted its son to serve his apprenticeship with Sulzer. That son happened to be Rudolf Diesel. Thus, it was that Sulzer became the first company ever to build a marine diesel engine, and the company continues to play a dominant role in marine diesel propulsion.

The company expanded into many other fields of heavy engineering, establishing subsidiaries all over the world. In the period 1933 to 1953, Sulzer developed a new type of weaving machine – today its most successful product, accounting for almost one quarter of its turnover – and we shall return to this later. Today the Sulzer group of companies employs over 30,000 people with sales of around $2.5 billion/year.

10.2. Why Natural Gas?

Like many large companies, Sulzer is organized into semi-autonomous business groups, each representing a particular product line. These develop their products and their markets without much regard for opportunities lying outside their defined field of activity. In order to enable the company to benefit from "extraterritorial" opportunities, two groups were formed outside the product divisions, one of them concerned with technical innovation and new products, represented here by Jack Metz, and another, similar, market-oriented group, represented by Tom Schur. The two groups work very closely together and it will readily be understood that, for a company like ours, looking at new opportunities in the energy field is an obvious requirement.

The great opportunities in the natural gas area came to Schur's attention through his acquaintance with Professor Thomas Gold of Cornell University (see Chapter 5 in this volume). The information Gold had accumulated all around the world appeared to indicate that the world's gas reserves were many times larger than the prophets of doom of the Club of Rome and their disciples would have had us believe. In 1983, as a member of the Panel on New and Renewable Energy Resources at the World Energy Conference in New Delhi, Schur stated: "I feel we must not reject depletable resources so categorically as is now fashionable to do". This was in sharp contrast to the inaugural speech by the late Mrs. Indira Gandhi, who still thought that our energy future depended in the main on solar energy and thought that photovoltaic devices should or could be produced at a sharply lower price.

10.3. The Swiss Energy Scene

Switzerland is a small, barren, land-locked country devoid of natural resources except for its hydropower, which has been fully developed but which contributes only about 15% of Switzerland's energy. The rest is imported, mainly in the form of oil. Swisspetrol, together with a major German exploration company, poured several hundred million dollars into oil and gas exploration but, so far, has achieved only modest success in the form of a very small gas well at Finsterwald near Lucerne.

Gold's theories, however, tell us that the geology on which the Swiss exploration is currently based is antiquated, if not antediluvian, and that Switzerland, in common with other regions of the world where the impact of two tectonic plates has created big mountain ranges, ought to have large reserves of abiogenic natural gas. Obviously, from the Swiss national point of view it would be of great importance to find an indigenous source of energy at almost any cost. At least the Swiss would know that they have such a strategic reserve even if, with their free-trade mentality, they continued to import energy from other competitive sources. Sulzer is interested not only in this strategic reserve, but in the opportunity to develop equipment specific for the methane age. An indigenous source of methane gas in Switzerland would give a tremendous impetus to such developments. It will readily be understood, therefore, that Sulzer intended to carry out the preliminary geologic exploration.

10.4. Tunnel Vision

The fact is that Sulzer has not yet initiated the preliminary geologic survey of parts of Switzerland, and this is due to the resistance of the exploration experts in Germany, whose skepticism about Gold's controversial theories sometimes seems to take on the form of holy war. We fully respect expert views, and we know that, in many cases, they know best and possess knowledge that is not available to the uninitiated; but we also know that there are cases when experts cannot break out of the structures defined by their traditional thinking. At this stage of our exploration endeavors, we are thus less restricted by technological problems, financing worries, or geologic uncertainties than by a phenomenon known as tunnel vision.

I have deliberately dwelt on this point, since tunnel vision is found not only in geologic circles but in almost all spheres of science and engineering. On the other hand, the engineer, if he is an innovator, has the opportunity to apply new knowledge or new technologies to traditional fields and thus challenge the complacent "expert", who is afflicted by such tunnel vision.

Let me raise the example of the Sulzer weaving machine mentioned earlier. Weaving was traditionally done by inserting a thread – the weft – from a spool in a shuttle. A German inventor came up with the idea of shuttleless weaving in the 1930s, when there was a strong Swiss textile machinery industry, of which Sulzer was not part. Sulzer, however, embarked on this development, which the "experts" considered to be impossible, and after 20 years of development work, put on the market what is now the dominant technology: the shuttleless weaving machine.

Gas in the form of town gas has been around for a very long time. The gas industry is well established and very traditional. It is our great hope that in each facet of this industry we can find opportunities that have not yet been clearly perceived, if we can view this situation from the outside and apply new technologies or discover new market opportunities.

10.5. Engineering Opportunities in the Methane Age

Figure 10.1 shows the various sectors of the gas industry; obviously, each will benefit if more natural gas at competitive prices emerges. The benefits can mean expanding markets for the traditional technologies as well as new markets through new technologies. To bring the full potential of these new opportunities to fruition requires an active interplay between those engaged in the market, in science, in geology, and in engineering, in such a forum as the one that generated this book.

As engineers, we can usually apply the best solutions only if we know where the market people or the scientists have encountered problems or reached technological frontiers not previously crossed. Every such problem can turn into a challenging task for us.

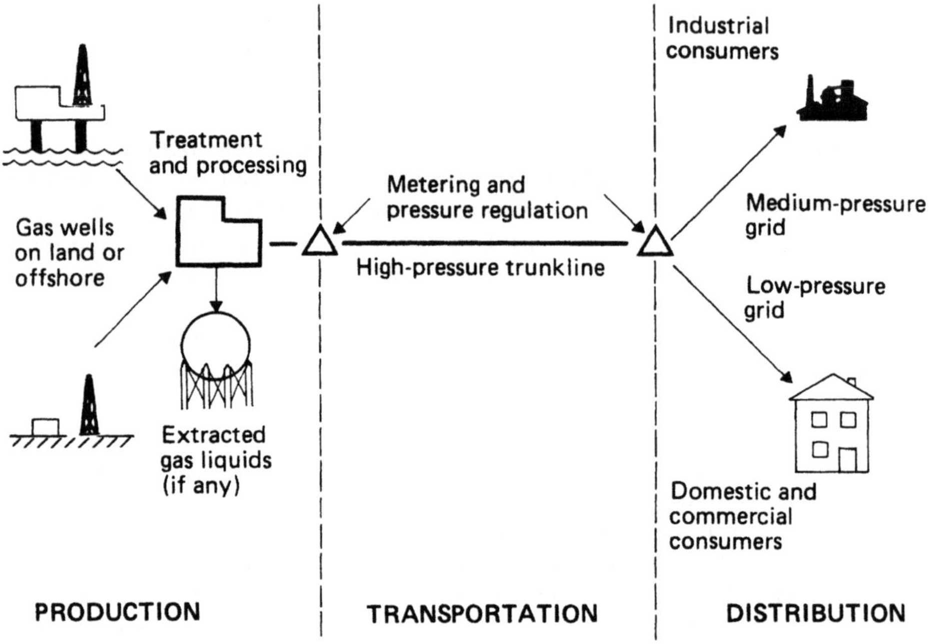

Figure 10.1. Natural gas supply – main activities.

We can start at the left of the diagram (*Figure 10.1*) and take, as an example, gas exploration. One opportunity here may be the application of improved geological techniques in order to achieve a higher hit rate. On the other hand, deep gas deposits may require new drilling methods into basement rock rather than into sediment. Alternatively, methods may have to be devised to release methane, which some think exists in large quantities below frozen methane hydrates deep down in the bed of the sea. The deployment of development funds into one or more of these areas is a matter for serious discussion and evaluation.

10.5.1. Making gas more competitive

After the "energy crisis" of the 1970s, with its associated psychosis of apparent "limits to growth", we have returned to the more healthy competition among the various energy resources for market share. In this competition availability, cost, and convenience, as well as compatibility with the environment, are the main criteria for the success of one resource against another. Here the engineer can do a lot to enhance the efficient use of gas. For instance, perhaps one of the most remarkable recent developments is the new condensation furnace for home heating purposes. Whereas previously heat escaped from domestic furnaces through the flue, in this new development all moisture is condensed out within the furnace, retaining its total heat. The completely cooled down and dried flue gases can then be safely expelled. This home furnace saves some 15 to 40% in energy costs over conventional gas furnaces and is a development not yet matched by the oil-burning alternative.

By contrast, little development work has yet been done on prime movers using gas, an area in which the large marine-type diesel engine, burning the heaviest residual oils, has reached remarkable efficiencies of up to 53%. *Figure 10.2* shows the historical development of efficiencies of different prime movers. First of all, the remarkable rise in recent years of the oil-burning marine-type diesel engine must be observed. The fact that this prime mover, burning the worst of residual heavy fuels, can now convert up to 53% of those fuels into mechanical energy has come about due to the fierce competition between two engine builders for the lead position in a declining market gripped by the fear of running out of oil. You will see that the gas turbine, the unit that thrives on being fed natural gas rather than crude oil or its distillates, has never reached anything close to that efficiency and, due to the difference in the thermodynamic cycles, will never quite catch up with the diesel engine. The comparison shows, however, that scope exists for improvements of efficiency. If methane enters into the competitive battle against other fuels, those who want to win that battle will have to ensure that the turbine gains a high rating in the overall energy cycle.

Efficiency in converting fuel to mechanical energy is, of course, not the only criterion, since even greater overall fuel utilization rates can be obtained if waste heat from the prime mover can be usefully employed. To get the best results will, however, require a dialogue between the engineer and the utility trying to expand its market. In fact, it is not yet certain whether it is worthwhile to spend much development money for high-efficiency gas-burning diesel engines when gas turbines, in the form of cogeneration units or combined plants, may be the answer for specific requirements. This may also be the case for newly emergent fuel cells.

10.5.2. Transportation

Until quite recently, we could not envision that the automobile could run on anything other than an oil-based fuel. With the emergence of natural gas in some parts of the world, conversion to compressed natural gas operation has been successfully introduced, and we are pleased that many CNG filling stations around the world are equipped with our compressors. These filling stations are, of course, copies of today's gasoline filling stations, similar to the way that the first automobiles were copies of carriages – horseless carriages.

Further development of natural gas use for motor transportation can go in several different directions. New and more economic methods to convert gas into a liquid fuel, such as methanol, may permit the existing gasoline-based infrastructure to be used. On the other hand, the

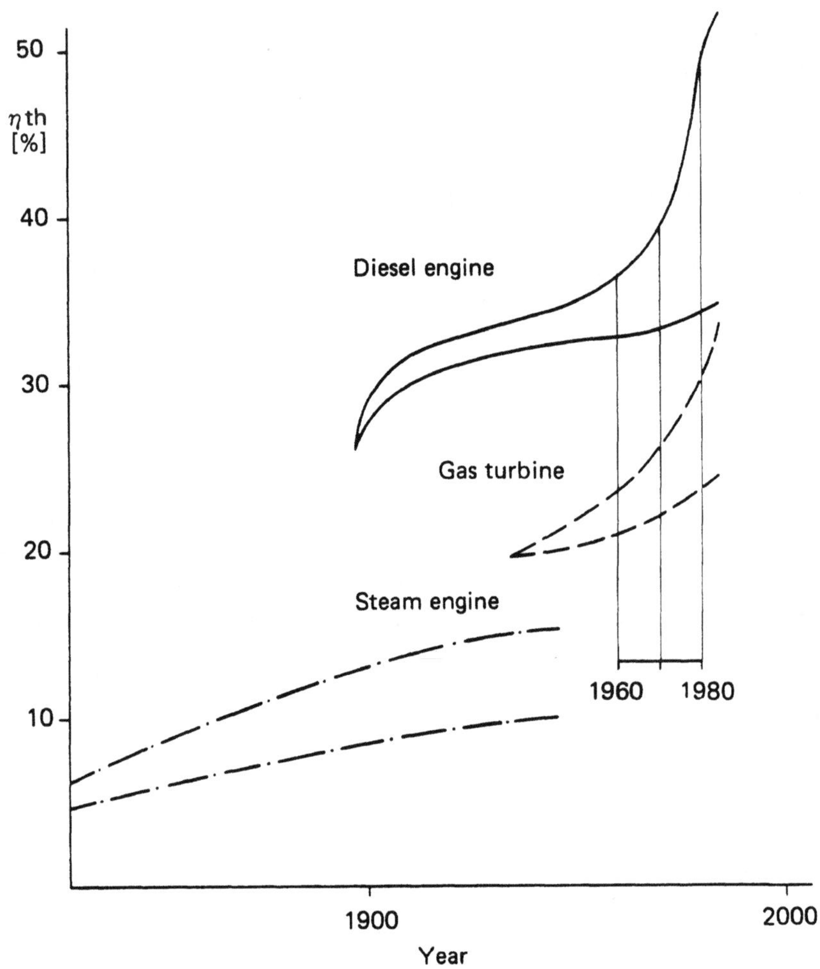

Figure 10.2. Historical development of prime mover efficiency.

convenience of gas in areas with a well developed gas distribution network may lead to the greater convenience of filling up a car overnight with CNG in one's own home. Whatever the route, whether the consumer will get his methane in liquid form or in compressed form, there are some challenging tasks before the engineer to prepare the necessary equipment.

10.6. Conclusion

New discoveries, new theories, and new technologies are continuously challenging our traditional notions. They arrive with such rapidity that they are difficult to absorb and are considered to be a threat by those who have successfully pursued their careers based on traditional assumptions. Our competitive world gives an advantage to those capable of understanding the new discovery and capable of interpreting the resulting trends.

The energy scene, so solidly based on coal at the start of the industrial revolution, and then again on oil, has recently undergone a most fundamental shakeup from which natural gas appears to be emerging as a leading source of energy. In this respect (as in many aspects of science and engineering), there are those who cannot keep pace with the change and who do not realize that we are experiencing something akin to the Copernican revolution in energy.

CHAPTER 11

Ecological Advantages of Natural Gas over Other Fossil Fuels

N. Chizhov and M. Styrikovich

11.1. Introduction

When estimating the total public costs of traditionally used energy resources one has to take into account, among other components, expenditures on environmental measures, which might be considered a function of the emission constraints per fuel consumption unit. There are also social costs in compensating for the damage from environmental pollution through discharges and exhausts of pollutants, plus residual damages for pollution derived from the technology as a whole, which affects all the components of the geosphere, as well as public production branches and elements. Most studies in this field are concerned with atmospheric pollution by ash, sulfur, and nitrogen oxides. These represent (including CO emission, which is negligible at electrical stations) over 90% of pollutant emissions.

These studies also take into account indirect pollution of hydrosphere, soil, and biosphere by precipitation of pollutants from the atmosphere (acid rain, etc.). For all that, problems connected with direct hydrospheric pollution by, for example, the washout of dissolved substances from ash dumps, soil pollution, "minor" pollutants, and the heavy metal aerosols and carcinogenic cyclic aromatic hydrocarbons are neglected.

It is a rather complicated problem to estimate the damage incurred by the biosphere and society from technologically derived pollution, since it requires performing a detailed study of the pollutant turnover in Nature, and the reaction of all recipients to contact with a pollutant, taking account of the contact intensity and its duration. Every particle of the chemical substance, while participating in biochemical cycles, can, when passing from one medium to another, exert a harmful influence for a long period of time. Therefore, the damage per unit (e.g., damage per 1 kg of technogenic SO_2)

will gradually increase. This process can continue till the harmful substance transforms into a harmless compound or takes up an inert form and becomes a deposit that does not take part in any active turnover (e.g., chemically inert oceanic bottoms). Mechanisms of pollutant migration, transformation, and deposition (nonparticipation in the active turnover) have not yet been sufficiently studied.

Over the last few years in the USSR and other countries, methods have been developed that allow the estimation of damage from pollution and its comparison with expenditures on its prevention, i.e., the "expenditure–effect" estimate. Based on these methods, an endeavor to estimate the damage caused by a hypothetical thermal station has been made. These studies, though incomplete and at a preliminary level, are worth thorough investigation, full support, and further development. The first part of this chapter deals with an attempt to compare the social losses suffered from the consumption of different forms of organic fuel. The estimates presented next are based on a review of some non-Soviet studies and additional estimates performed by the present authors using data from the studies mentioned.

11.2. The Social Costs of Pollution

Temporary methods have been established for estimating the cost of countermeasures against thermal stations and big boiler house discharges. These methods, confirmed by the USSR Energy and Electric Power Ministry in 1982, evaluate the economic efficiency of environmental protection measures by means of the minimum-cost or payback-term criterion. One should take into account changes in technical and economic indicators in this evaluation, such as decrease in the losses suffered by the national economy from atmospheric pollution, and in some cases the benefit gained from additional products derived from trapped substances. The total economic effect represents a cumulative saving of annual national economic expenditures allocated to energy, public health, municipal services, agriculture, and forestry, which are determined in correlating the costs for the base case and the one under consideration. A correlation with ecological consequences is assumed.

For all that, one requirement should be observed: a level of the near-ground atmospheric layer pollution, expressed in conventionally modified units, which does not exceed standard public health norms. Such a requirement governs the choice of technical solutions. Provided the capacity of a given enterprise and the quality of the burned fuel remain constant, the cases are ecologically comparable, if the product of generalized indices of the dispersal conditions per volume of the pollutant discharge remains equal.

Developments outside the USSR in this field, using thermal stations, as an example, are generalized to a certain degree in Shang-zhi Wu's (1984) study, which is mostly based on the methods used in the research efforts accomplished under the aegis of the OECD (1984). Awad and Vezirogly (1984) in their study attempted to estimate the damage per tonne of coal equivalent (tce) on a global scale, taking account of long-distance transfer, deriving damage estimates per component. However, all these estimates were performed with no thorough scientific studies on pollutant impact upon the environment at each stage of pollutant migration; thus, the insufficiently developed routines for summing the damage per unit and by component should be considered rather approximate. The above estimates turn out to yield order of magnitude figures for the phenomenon under consideration, rather than a precise quantitative estimate.

Wu's paper discusses a new (projected) thermal station running on coal with a rated capacity ranging from 500 up to 2,000 MW as an example when estimating an ecological component within the total costs of electric energy. All costs are in 1982 US dollars; the constraints on ash and SO_2 discharges are in accordance with those adopted in the USA and countries of Western Europe. The average expenditures on atmospheric protection from dirty gases of thermal stations generated by coal are estimated at $250/kW of capital investment and about of 1 cent/kW/hour of annual operating expenses.

The techniques and technology of today cannot provide protection for the atmosphere from thermal station pollution to a greater degree. In this connection, the estimate of the expense needed to meet the losses suffered from such pollution plays a significant role.

According to Wu, the OECD attempted to estimate the damage from thermal station discharges of SO_2 into the atmosphere and the damage reduction resultant from declining exhaust emissions. The economic damage (the first component) in this endeavor is determined as the sum of the damage incurred to structures from corrosion, decreasing crop capacity, and decline of the fishing industry, and leaves aside the damage inflicted upon forestry, cattle breeding, cultural and historical values, recreational landscapes, etc.

The harm to public health (the second component) has been determined separately and interpreted, eventually, as the loss of production resulting from a general decrease in the mean life expectancy. It remains unclear whether this estimate also covers increased expenditures on public health, social security, and the labor productivity decrease due to the deterioration of individual health. It is also unclear whether both the loss of production due to a faster attrition rate of the means of production plus additional depreciation expenses, or only the latter, were taken into account. There is no estimate of the damage suffered from all the pollutants contained in all the thermal station discharges, either. It is evident, therefore, that all the above estimates are only preliminary, failing to reflect to any complete degree an "ecological" component in the total social costs of the energy produced by thermal stations.

In estimating the benefit from a decrease in pollution per country, the transboundary migration was assumed to be conventionally equal to zero. A linear dependence of damage decrease on reduction of pollution was assumed as well. The study presents a wide range of estimates of both the first and second component in the SO_2 damage. For example, at a discharge of 7.5 kg of oxides per tce the first component (economic damage) is estimated at from $1.60 up to $9.75 per tce or $0.21–1.30 per kg of SO_2. At the 15 kg/tce discharge level, the estimate is $3.25–16.25/tce, or $.21–1.08/kg of SO_2. At the same time, these estimates differ greatly over various countries. Low estimates range from $.04/kg of SO_2 up to $2/kg; the top ones range from $.21/kg up to $22.50/kg. On average, for 11 countries of Northern and Western Europe, a low estimate is about $.25/kg, the top one being $2.22–2.40/kg. The estimates for West Germany are $.4/kg and $45.50/kg, respectively.

The second (social) component, in different versions, is estimated at from $7 up to $150/kg and from $6 up to $266/kg for the expected life duration. Since the expenses for decreasing discharge represent annual estimates, it is only logical, in order to obtain an annual estimate benefit, to divide these figures by the expected duration of life . Assuming the latter as being equal to 70 years, we obtain $.10–2.10/kg per annum and $.09–3.80/kg per annum. Provided that the economic and social damage, on average, is of the order of $2/kg, and given a discharge of 15 kg/tce, the damage per unit from the atmospheric pollution by SO_2 from thermal stations using coal is estimated at $60/tce. This is only a partial estimate, accounting for the SO_2 discharge into the atmosphere; the total damage will be about $180/tce. The latter estimate is of the same order as the ones presented next, which are calculated by the present authors using the data of Awad and Vezirogly (1984) and *Options* magazine (1984).

The average net annual benefit (effect minus expenses) from the pollution decrease when summing the first and second components, and setting the discharge decrease at 4.8 million t/year or 19%, is estimated at $34.60/head and, in the case of the entire population, about $8 billion/year; having the discharges decrease by 11.7 million t/year or 48%, we get $98.60/head and $16.5 billion/year, respectively. If one extrapolates the expenses/losses ratio, the total benefit from the above reduction in cumulative biospheric burden from indigenous sources can be estimated at $50 billion/year or $200/head/year.

Awad and Vezirogly estimated the global damage from 1 tce of organic fuel. When correcting the data, the present authors estimate the damage from 1 tce of hard and sulfurous liquid fuel at about $156/tce. The damage from gaseous fuel per unit of the primary energy consumed was assumed by these scientists to be three times lower than that from the hard and liquid fuels. Under such an estimate, the damage from natural gas on average will be $52/tce and the estimation costs of the benefit when replacing coal and fuel oil by natural gas is $100/tce. Based on the data mentioned in this paper, the present authors calculated the damage per unit from different organic fuels, taking into account the total energy consumption structure (*Table 11.1*).

Table 11.1. Estimate of the damage per unit to society from thermal station use of organic fuel (in 1984 US $ per 1 tce).

	Type of fuel		
Environmental component	*Solid*	*Liquid*	*Gas*
1. Population	38	38	13
2. Stock of land	15	-	-
3. Agriculture and forestry including:	29	29	9
plant growing	12	12	4
cattle breeding	5	5	1
forestry, fauna, and flora	12	12	4
4. Fishing resources and industry	3	3	1
5. Buildings and structures	32	30	13
6. World ocean coast	-	6	-
Total[a]	117/107	106/92	36/30

[a]Numerator = damage to coastal countries; denominator = damage to inland countries.

Damage to coastal countries includes the damage to the sea coast through oil and oil product leakage; damage resulting from a change in the ocean level due to a probable change of the climate seems to be negligible at least until 2050.

By these estimates, the damage decrease from replacing hard and liquid fuels with natural gas is $60–80/tce. Our study also contains estimates of the average damage from synthetic gas (similar to natural gas in its composition) equal to $150/tce, and from synthetic fuel oil at $217/tce. The damage decrease with the substitution of residual liquid fuel for gas is $70/tce in our case.

Using data presented in *Options* magazine, the present authors attempted to estimate the approximate damage per unit for the USA, which turned out to be $190/tce.

Therefore, the benefit for developed countries (the decrease of damage to society) to be gained by replacing coal and heavy fuel oil with natural gas in the energy balance, as a first approximation, might be from $60 up to $120/tce. Preliminary estimations show that such a benefit might be about $80/tce (in 1984 dollars), which shows that the use of ecologically clean fuels (notably, natural gas) provides additional gains within the range mentioned.

An assessment of energy resources with regard to social costs makes one see the effectiveness of individual technologies in quite a different light. In this respect, technologies based on natural gas have substantial economic advantages over other fuels.

In the calculations that follow, this aspect has not been considered to any fuller degree. Therefore, the willingness-to-pay estimates for natural gas are performed proceeding from the correlation of "explicit" economic consequences when consuming gas and other alternate energy carriers. In fact, a substantial ecological component of the effect ("implicit effect") might disturb the market equilibrium, causing higher gas consumption, and thus promoting an increase in the selling price. The GRI baseline (1984) predicted that price differences (for the USA) between natural gas and low-sulfur coal would be about US 1985 $80 per tce in 2000 and over $100/tce after 2000. A considerable increase of the social value of gas over its selling price will allow governments to tax other, "dirtier" energy carriers and thereby induce manufacturers and consumers of gas to extend the gas market.

Due to the lack of information on the economic damage resulting from the burning of various fuels, this chapter takes into account only the expenditures involved in environmental protection (notably, stack gas cleaning). Based on the analysis of Soviet and foreign publications, rough cost estimates have been made of desulfurization for the most widespread technologies of fuel burning (*Table 11.2*). The estimates are given for the periods prior to and beyond the year 2000; it is assumed that in the future the application of new, cheap, and more efficient methods of gas cleaning will probably lower the costs. But even after efficient cleanings, stack gas from today's electrical station based on coal is not so clean as natural gas emissions, because some part of SO_2 (10–15%) remains in the exhaust, as well as some solid particles, heavy metal aerosols, etc. These have been taken into account in the calculation of marginal natural gas prices.

Table 11.2. Additional expenditures involved in stack gas cleaning and the burning of solid and liquid fuels compared with natural gas ($/tce).

Fuel type	*Prior to 2000*	*Beyond 2000*
Solid fuel (low-sulfur coal):		
Thermal power plants		
base-load	30	20
intermediate-load	45	30
Heating boiler plants	55	20
Industrial boiler plants	45	20
Liquid fuel:		
Thermal power plants		
base-load	15	15
intermediate-load	20	10
Heating boiler plants	30	15
Industrial boiler plants	20	20
Household installations	15	15

Figures in *Table 11.2* represent costs for desulfurization only. If we included $DeNO_x$ process expenses, those costs would be much higher today. Of course, $DeNO_x$ also exists in natural gas combustion products, but this fuel has no nitrogen bound in molecules of the fuel itself, which form $DeNO_x$ even at moderate temperatures

Natural gas burning creates $DeNO_x$ by oxidizing atmospheric nitrogen, which takes place at very high temperature only. Such formation can be reduced by simple changes of combustion conditions. All new stationary installations, of big or moderate size, must include as a first step gas turbines in which burning takes place at a very high air excess ratio. In such cases, even maximum theoretical temperatures can be lowered drastically and NO_x formation reduced to the level near to concentration of NO_x at coal stations equipped with $DeNO_x$ installations.

Today $DeNO_x$ is very expensive, and forecasting its use at today's price would be unreasonable. But in future, even cheaper $DeNO_x$ installations will still make the burning of coal significantly more expensive than natural gas.

Another factor must be taken into account: the qualitatively higher efficiency of natural gas installations in electric stations, especially in cogenerating systems, which leads to smaller fuel consumption and, consequently, to decreased environmental damage. In addition, CO_2 emission from natural gas, even at the same efficiency, is about half of coal's (per tce), which reduces the possibility of climatic impacts.

Of course, all existing estimations of the ecological advantages of natural gas are only preliminary and more or less valid for big consumers only. But even these preliminary figures emphasize the major significance of the problem and the necessity for more profound investigations. Such research must cover the specific conditions of various areas, especially developing countries, and the use of natural gas by different consumers.

References

Awad, A.H. and Vezirogly, T.N. (1984), Hydrogen versus synthetic fossil fuels, *Int. J. Hydrogen Energy* **9**(5), 355–366.

OECD (1984), *The Costs and Benefits of Sulphur Oxide Control* (Organization for Economic Cooperation and Development, Paris).

Options (1984), Number 1 (International Institute for Applied Systems Analysis, Laxenburg, Austria).

USSR Energy and Electric Power Ministry (1982), Temporary methods for estimating the economic efficiency of measures aimed at atmosphere protection from pollutant discharges containing dirty gas of thermal stations and boiler houses (Soyuztechenergo. M., Moscow).

Wu, S. (1984), *Air Pollution Costs of Electric Power Generation Systems* (Massachusetts Institute of Technology, Cambridge, MA).

CHAPTER 12

The West European Natural Gas Market

Måns Lönnroth

12.1. Introduction

I once read that power is as fundamental to the understanding of society as energy is to the understanding of nature. My task is to try to understand the relationship between energy and power in this societal sense. Taking IIASA's Technology–Economy–Society (TES) Program as a starting point, my interest is really in the *technology's* links to *society*. The *economics* in TES is, to me, a tool for understanding this interrelationship – not a very good tool at that. Economics handles bookkeeping between technology and society, rather than being a subject in its own right.

The fundamental thing to understand about natural gas in Western Europe is the interrelationship between the huge investments required, the market share of natural gas, and the various institutional structures. One such investment is the Norwegian Troll field; others of similar magnitude are the Siberian pipeline, the Algerian Trans-Med pipeline, and the LNG transportation network from Algeria to Western Europe.

My initial main interest was in the competition between the Soviet Union and Norway for the West European natural gas market, as a background for the Swedish discussion on nuclear energy and natural gas. This led me to complementary perspectives. For instance, the essence of Rogner's Chapter 4 in this volume is that none of the gas studies he had looked at mentioned technology. Price, profits, etc., are frequently mentioned, but never technology. I noticed that power is also not mentioned, so I take that as my starting point.

With respect to the Troll field, there are two elements to note: first, it forms 25% of Western European gas reserves, which is huge; second, it is located in over 300 m of water, within a complex geology, and the estimated cost for one platform is perhaps $5–6 billion. Compared to the amount of energy produced annually, it is still considerably cheaper than, for example, nuclear energy. Moreover, Norway is a country of 4.5 million inhabitants on the periphery of Western Europe. The project has been called a Norwegian "man on the moon project" because of its impact on Norwegian industry, technology, and skills. Varnado's comment, in Chapter 6, on the technology leaders and the role of Statoil, are encouraging in this regard.

12.2. West European Scenarios to 2000

The West European natural gas market, as is well known, increased rather dramatically from 1960 to 1980; since then, it has by and large stablized. Production inside the EEC peaked in 1975 and then gradually declined. Projections for the year 2000 suggest that total imports to the EEC may fall somewhere between 60–130 Mtoe. Contracts in early 1986 would yield 55 Mtoe (Algeria 25, Soviet Union 25, Norway 5) at the turn of the century. Most of Norway's current contracts, however, will terminate in the 1990s. Contracts from the Soviet Union and Algeria could easily be prolonged, given mutual goodwill between exporters and importers. Thus, the total need in the West European market for entirely new import contracts could range from 0 to 50 Mtoe. The recent contract for Troll–Sleipner gas covers roughly 20 Mtoe.

The biggest uncertainty is the UK, where the situation may involve a certain amount of exports to rather substantial imports, depending upon various assumptions about the market and UKCS (United Kingdom Continental Shelf) production profiles. West Germany could well require substantial imports, but the actual amount will depend on the competitive situation. In the 1990s gas will compete not only with oil but also with other fuels, such as coal (given the overcapacity of the coal industry) and electricity (given the planned nuclear power additions, which in turn will displace coal). Competition between gas and electricity also remains strong in other countries – France, most notably, but also Great Britain and Belgium. In Denmark, gas competes with district heating.

It is important to note that the market share of gas in Western Europe is to a large extent constrained. In the FRG, marketing is essentially self-restrained, to maintain the consensus of what are reasonable market shares among electricity (and nuclear power), coal, oil, and gas. In France, the restraint is essentially coerced; and Gaz de France is sometimes called the little sister of Electricité de France. The situation in the UK is similar, but gas holds a stronger position there.

12.3. The Post-2000 Competition

Beyond 2000, the interesting competition will be between gas and electricity, and more directly between gas and nuclear energy. There is evidence of this already in some countries. The West German case is largely driven by nuclear energy, since its introduction displaces coal, which in turn is guarded by an agreement between the coal industry, the electricity industry, and the government. The net effect is to restrain the market for gas.

Let us turn to trade structure. Of the four sellers – Algeria, the Soviet Union, Norway, and Holland – three selling companies are state-owned; the fourth, Gasunie, is 50% state-owned. Among the buyers British Gas, SNAM, GdF are state-owned, while the role of the state in Ruhrgas is less direct. (It is too early to fathom the implications of the privatization of British gas.) The German situation, with essentially private ownership, is perhaps the most interesting, but I will defer my analysis of that case to another occasion.

The state is not equally influential in all countries. The French government is a different animal from the Italian. In some countries the companies make state policy rather than the other way around. How consensus on the different perspectives is formed is the important issue.

12.3.1. Shell Oil

It is worth devoting a few words to Shell, which is widely agreed to be the most influential oil company in Western Europe. As a gas company, it is part owner of Distrigaz, Gasunie, Ruhrgas, and DUC, as well as some other companies in the FRG. It is an operator in the North Sea in the British, Danish, and Norwegian sectors. It acts very much as a communications switchboard and as a marriage broker.

Shell and Esso Europe work together in two of the most important countries: they are equal partners in Gasunie and Ruhrgas. Esso left their partnership in Distrigaz, and they are not partners in the North Sea. People within the industry seem to perceive Exxon as being very much an American company – monolithic, centralized, run from the USA. Shell is more a federation of companies, with a looser structure, but with a clear sense of common purpose. The federated structure is part of the reason why Shell is capable of playing a more diverse role than Exxon.

12.3.2. Ruhrgas

The most important natural gas company in Western Europe is Ruhrgas. It is central because of its technical competence, its geographic position, the size of its market – the biggest on the continent – and because of its influence within West Germany. Essentially, Ruhrgas is a wholesale company playing a dominant role within the West German natural gas industry due to both the legal structure (the so-called *Demarkationsvertrag*) and the ownership structure. This is a demonstration of power balancing in its own right.

Ruhrgas also plays an important role in East–West trade and *Ostpolitik*. It is without any doubt a commercial company, but it is also an expression of German national interest. There has been much talk, not least in the USA, about the FRG being dependent on Soviet natural gas; an equally important issue is the Soviet Union's dependence on hard currency income. Assuming that Soviet oil exports will decline toward the end of the century, its main hard currency earner will be natural gas, and the main marketer of this gas in Western Europe will be Ruhrgas.

12.3.3. Trade

The level of trade will depend on a mix of commercial and noncommercial factors. Starting with Soviet exports to Austria, it is fair to say that the Austrian view of itself as an East–West broker makes it a rather natural customer for Soviet gas. In the FRG, East–West trade and *Ostpolitik* are obviously linked. In Italy, counter trade is probably more important, given its capacity to export both pipes and other equipment. In France, again, it is not possible to distinguish between counter trade and foreign policy in general. Historical relations with France will give Algerian gas a role there. More generally, given the population increase in North Africa and the role of immigrants in the Southern European countries, there will always be an Algerian–French gas link, as well as an Algerian–Italian and an Algerian–Spanish relationship. Thus, both the Soviet Union and Algeria have what amounts to a reasonably assured share of the West European natural gas market, barring a sudden, severe political crisis.

Norway is a different case. Given the cost structure of its major undeveloped gas reserves, it is mainly geopolitical considerations that give Norway a future in the gas market. Its role is that of the balancer. Norway is the complementary factor, in particular, to Algerian and Soviet gas. Its main competitor is Dutch gas, for as long as the Dutch have some gas to export.

Returning to the Troll field, the market for 15 Mtoe of natural gas (or, for that matter, the 20 Bcm presently under discussion from Troll plus Sleipner) is by no means certain. Room can be made for these amounts, however, if other suppliers are squeezed to some extent, which is possible, given the flexibility clauses in the contracts. Ruhrgas can definitely make room for Norwegian gas if Dutch and Soviet contracts are squeezed. For

France, there is the possibility of not extending its Soviet contracts. Should the market expand toward the 230 Mtoe level, there will be no problems in accommodating the additional Norwegian contracts.

Ruhrgas's key position on the continental pipeline network gives it a unique potential for balancing Norwegian and Soviet gas imports into Western Europe. The Norwegian concept of bypassing Ruhrgas by building a pipeline directly to Zeebrugge is, in a way, a tribute to this position of Ruhrgas.

From the Norwegian point of view, new contracts covering its Troll production are essential, given the loading of the offshore industry. The big economic uncertainty is the short-term cash flow. An oil price of $20 per barrel may well be needed, but given the project lead times, it is likely that the price will be reasonable when the platform is in place. Much will depend on the rate at which Ruhrgas and Gaz de France may be willing to squeeze the Dutch and the Soviet exporters in order to make room for Norwegian gas.

12.4. Potential Markets

To return to the relationship between investments and markets, it is probably true that the investment sums relevant in the West European discussions require equally sizable actors. The commercial, political, and strategic aspects are too interwoven for smaller, private actors to be capable of sorting them out. This factor also has some implications for the market share of natural gas, which will only increase if the large actors find this to be in their interest.

This means, probably, a decoupling between gas and oil. Rogner states, in Chapter 4, that gas always has been a stepchild of oil. It is certainly true that in terms of power, the gas industry is a stepsister of the oil industry. If the two industries were decoupled and gas stood alone, it would be a rather weak industry. To increase the market share above the present roughly 15% level to perhaps 20% or 25% would require entirely new markets, given the West European situation. The only two available alternatives are the transportation and electricity sectors.

The transportation market is a technically feasible alternative, given technologies for gas compression. This would, however, bring gas squarely into the premium end of the oil industry market.

The second market, electricity, is ripe for some rather fundamental transformations, wedded as it is to the concept of gigawatt cathedrals. If electricity planners were to think about alternatives to coal and nuclear, such as combined cycle power plants, cogeneration on both small- and medium-scale levels, and so on, the outlook for gas would be different. This is the kind of ally the gas industry would need in order to increase its market share.

It would also be an interesting task for IIASA to look at gas and electricity both in terms of technology and in terms of institutional structure. Although electric utilities tend to be conservative, the industry is not completely monolithic. RWE, CEGB, EdF are, to be sure, large companies with a common view, but there are others. I would look into the West German situation and examine particularly regions where there is a multitude of institutional arrangements. Denmark is another interesting case, where the distributors have a relatively strong standing. Sweden might also present some opportunities, as well as Holland and some selected US utility companies.

An alliance with electricity is what gas needs in order to become independent of oil.

CHAPTER 13

Natural Gas and Hungary

D. Csako and J. Subai

This chapter is intended to give a brief survey of the supply and exploitation of natural gas in Hungary.

13.1. Past Production

The year 1987 marks the fiftieth anniversary of the beginning of the exploitation of natural gas in Hungary. Throughout this entire period, natural gas produced in Hungary has been a significant factor in the country's hydrocarbon energy supply (see *Table 13.1*).

Table 13.1. Hungarian domestic hydrocarbon production, 1937–1985

Energy source	*Quantity*	*PJ (approx.)*
Crude oil	64,280 kt	2,635.5
Gross natural gas	117,437 Mm^3	3,922.4
Gas products	8,925 kt	428.4
Total		6,986.3

This significant amount of domestic energy has been exploited through very hard, careful research and production–technological work requiring great expertise. In the course of this program 56% of Hungary's original potential geological hydrocarbon wealth has been explored.

The following conclusions can be drawn, based on the production history to date:

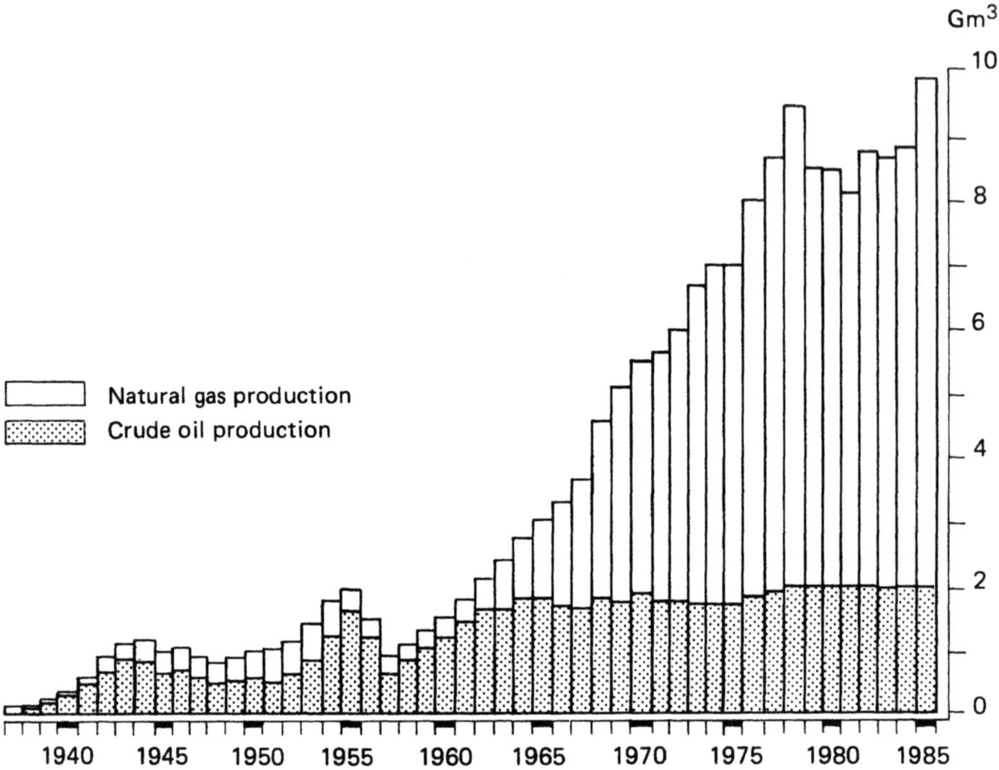

Figure 13.1. Hungarian crude oil and natural gas production, 1937–1985.

(1) The hydrocarbon reservoirs comparatively easy to explore have mostly been found; moreover, their exploitation has reached its final phase. *Figure 13.1* presents the time curve of Hungary's natural gas production.

(2) In the future, exploration will have to shift toward deeper, more complex reservoirs, and we expect an increase in the proportion of gas. This is expressed in the perspectives presented in *Figure 13.2*.

(3) Taking account of existing reserves and reserves expected to be disclosed, it is generally believed that domestic production will decrease (see *Figure 13.2*), and consequently hydrocarbon imports will increase in the future. This forecast presents special tasks, especially in the case of natural gas and particularly in terms of peak consumption management and related facilities.

(4) In domestic production we shall have to rely more on gas reserves with greater inert content – the production, pretransport preparation, and utilization of which will raise special problems to be solved.

(5) Another special technical task will be to find economic ways of exploiting low-yield gas fields.

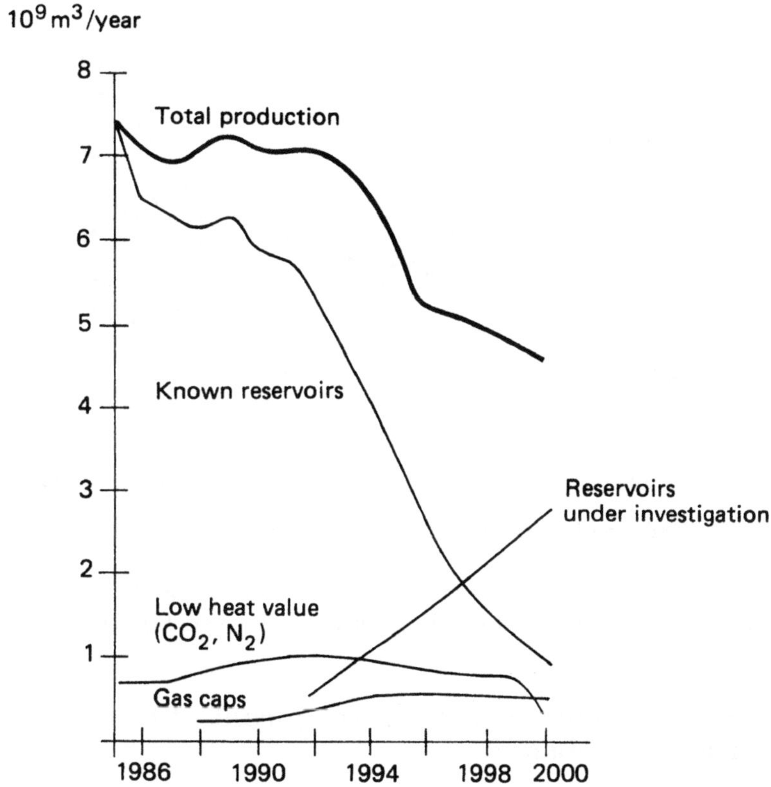

Figure 13.2. Planned natural gas production, 1986–2000.

13.2. Surveying Future Possibilities

Hungary's future hydrocarbon exploitation possibilities are thus basically limited by the predicted reserves.

Of the 93,000 km^2 of Hungary's territory, some 84% (approximately 78,000 km^2) is likely to contain hydrocarbon reserves. This promising part of the country can be divided into 13 basin areas from the exploration standpoint. Within each basin area we differentiate between Neogene, Paleogene, and Preneogene formation groups. If the potential wealth of the formation groups is regarded as 100 units, the ratio among the individual formation groups is expected to be 33–54–68%, respectively.

Considering the depth of the fields, we differentiate between the following categories: 0–1,500 m, 1,500–3,000 m, and below 3,000 m. The newly disclosed fields are expected to be below 3,000 m, such as Precambrian cracked metamorphites; Precambrian early Paleozoic and Mesozoic cracked caverned basins on the upper part of the bottoms of recesses; cracked

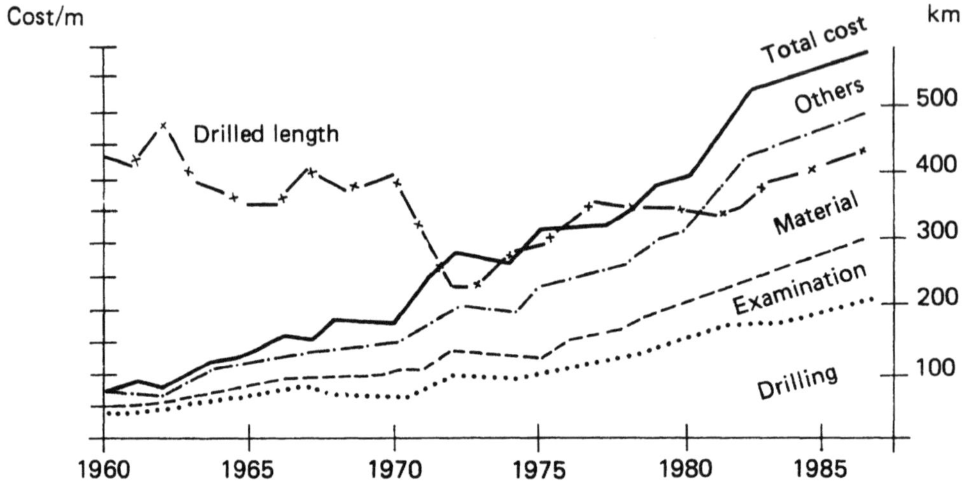

Figure 13.3. Drilled depth and specific costs, 1960–1985.

recesses formed inside Mesozoic carbonates; and Miocene detrital rocks. All this will involve higher well-drilling expenses – see *Figure 13.3*.

13.3. General Tendencies of the Development of the Natural Gas Industry

To interpret these trends, we should be familiar with Hungarian practices related to the production of gas and gas products. *Figure 13.4* describes these practices. It presents the logical–technological order that strictly defines the natural gas industry as a complex circle of tasks. The exploration activity for natural gas thus interpreted is organically and closely related to the most complex system of crude oil and natural gas production, presented on *Figure 13.5*.

Taking all that into account, the perspectives of the natural gas industry are influenced by three main factors:

(1) Absolute or relative energy shortage in the given country.
(2) Contemporary energy management and related technological level.
(3) The increasing role of governments in solving energy problems.

Regarding installations, the development goals are as follows:

(1) Greatest possible efficiency with low requirement in volume and space.
(2) Complex remote control, with as few instruments as possible and high operational safety.

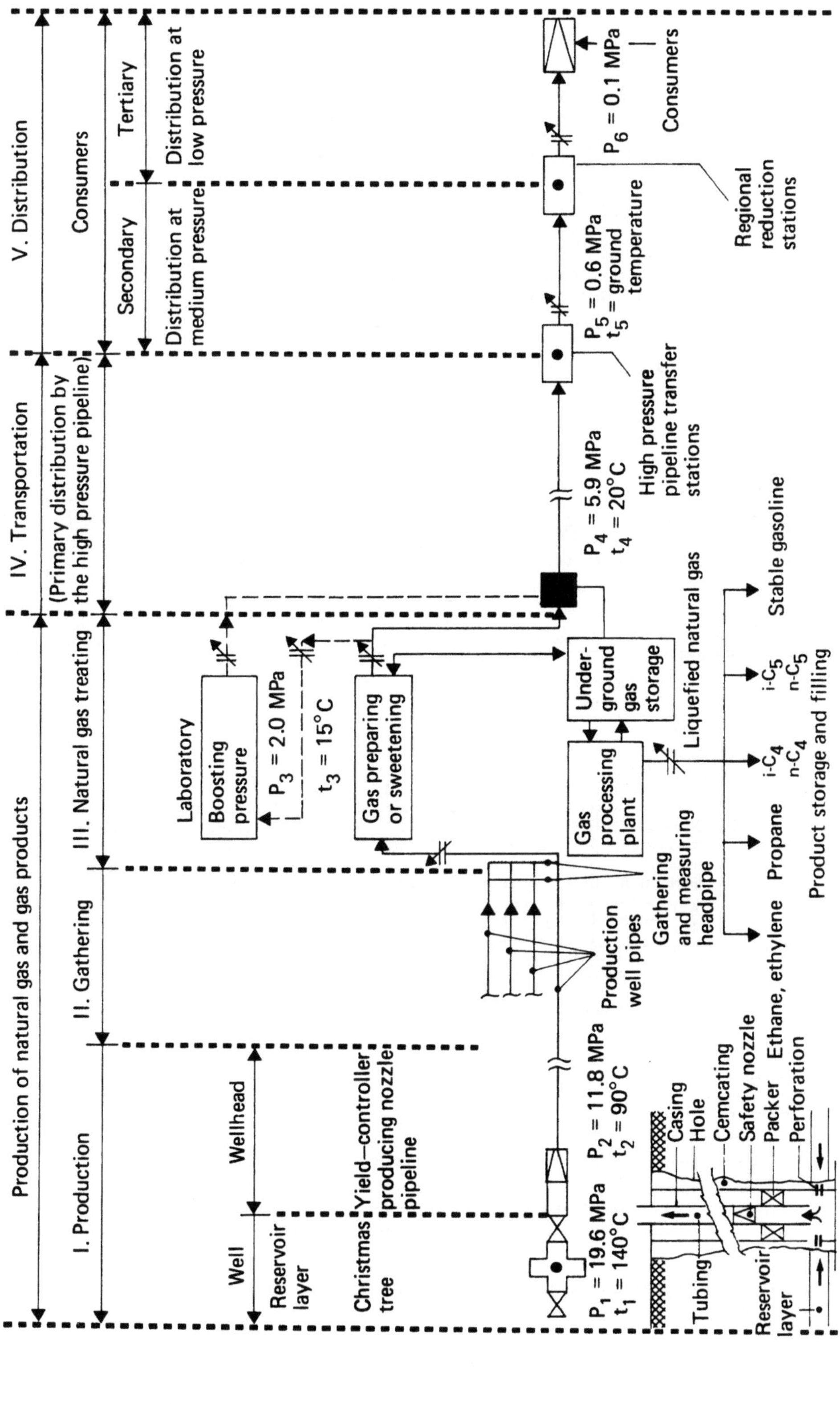

Figure 13.4. The route of natural gas from reservoir layer to consumer.

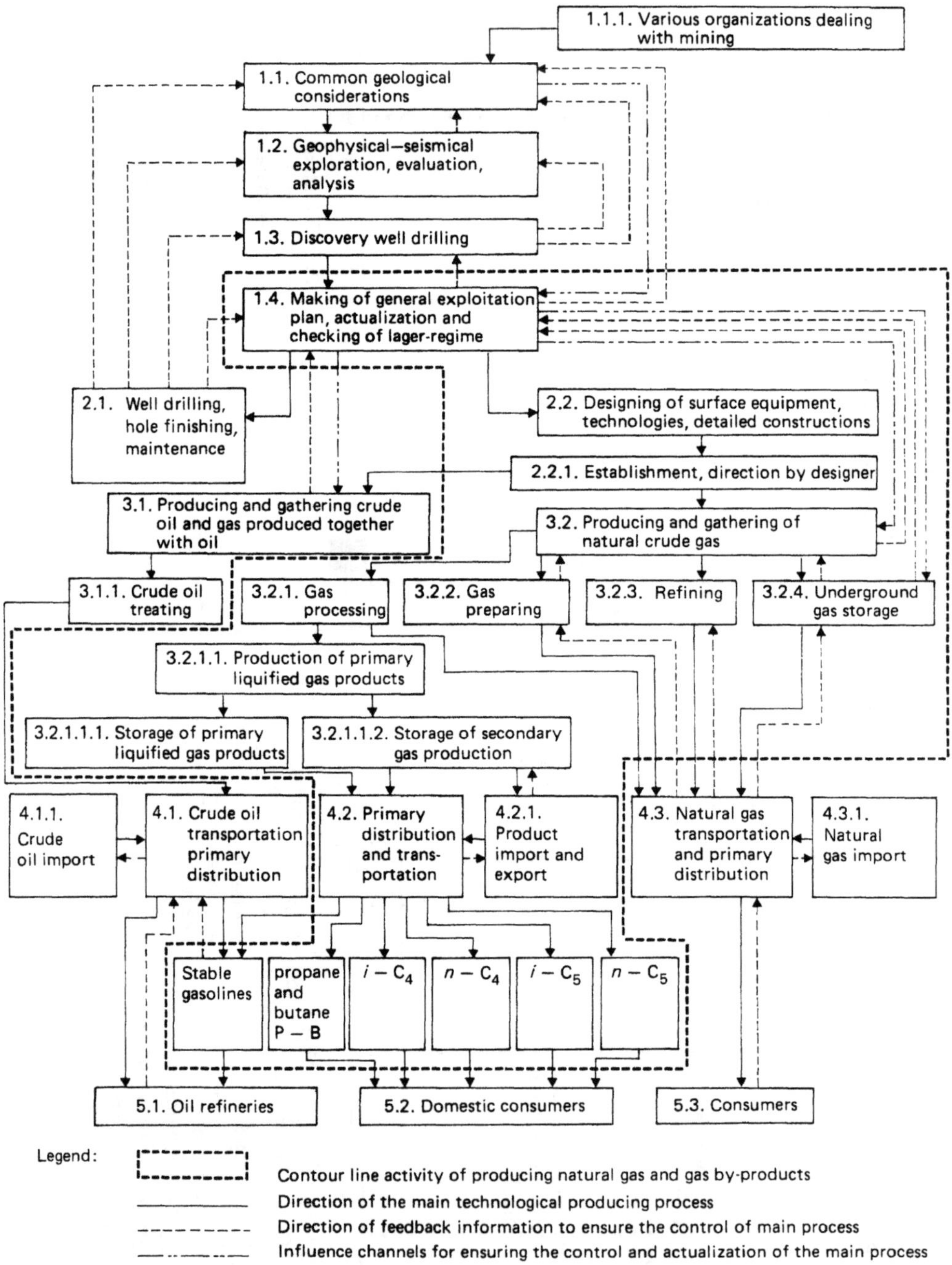

Figure 13.5. Connection of the natural gas and gas products industry to the hydrocarbon mining industry.

(3) Simple and reliable means to implement increasingly complex tasks.
(4) Increased exploitation and productivity with minimum costs.
(5) Lowest possible energy requirements for increasing production.
(6) Exploration of all possible resources and their exploitation with the maximum possible protection of the environment.
(7) Continuous operation with maximum safety and minimum labor.

Resolving these contradictory goals is a most interesting task in itself. The goals enumerated, which do not cover the entire picture, involve consequences worth mentioning:

(a) The regulations concerning environmental protection are becoming stricter every year, demanding special conditions for the systems producing, transporting, and utilizing hydrocarbons.
(b) Demands are increasing for secondary and tertiary production facilities.
(c) The role of inert gases is increasing.
(d) The efficiency of energy utilization installations has been significantly growing.
(e) New technologies and installations are being developed to utilize various energy wastes.
(f) Increased efforts are being made to reduce the effects of corrosion to a minimum.
(g) The safety level is significantly improving.
(h) Processes and technologies that pollute the environment are gradually being closed down.
(i) The operational safety of remote supervision, data transfer, control, and communications management installations is improving.
(j) The role of industrial robots is rapidly increasing.
(k) Given the time and space gaps between production and utilization, special transportation and storage problems must be solved.
(l) The ratio of communal to household consumers is on the increase, causing consumption to fluctuate, which has to be balanced by special measures.

All these factors also have to be dealt with in Hungarian practice.

13.4. The Role of the Natural Gas Industry in Hungarian Energy Management

The natural gas industry has played a significant role in Hungarian energy management. Within this industry the exploitation of domestic wells is of decisive importance. In addition to producing and delivering natural gas, this management activity includes the manufacture of gas products; meeting various consumers' interests; peak-balancing technologies; high-pressure, long-distance transportation; and primary management operations with gas

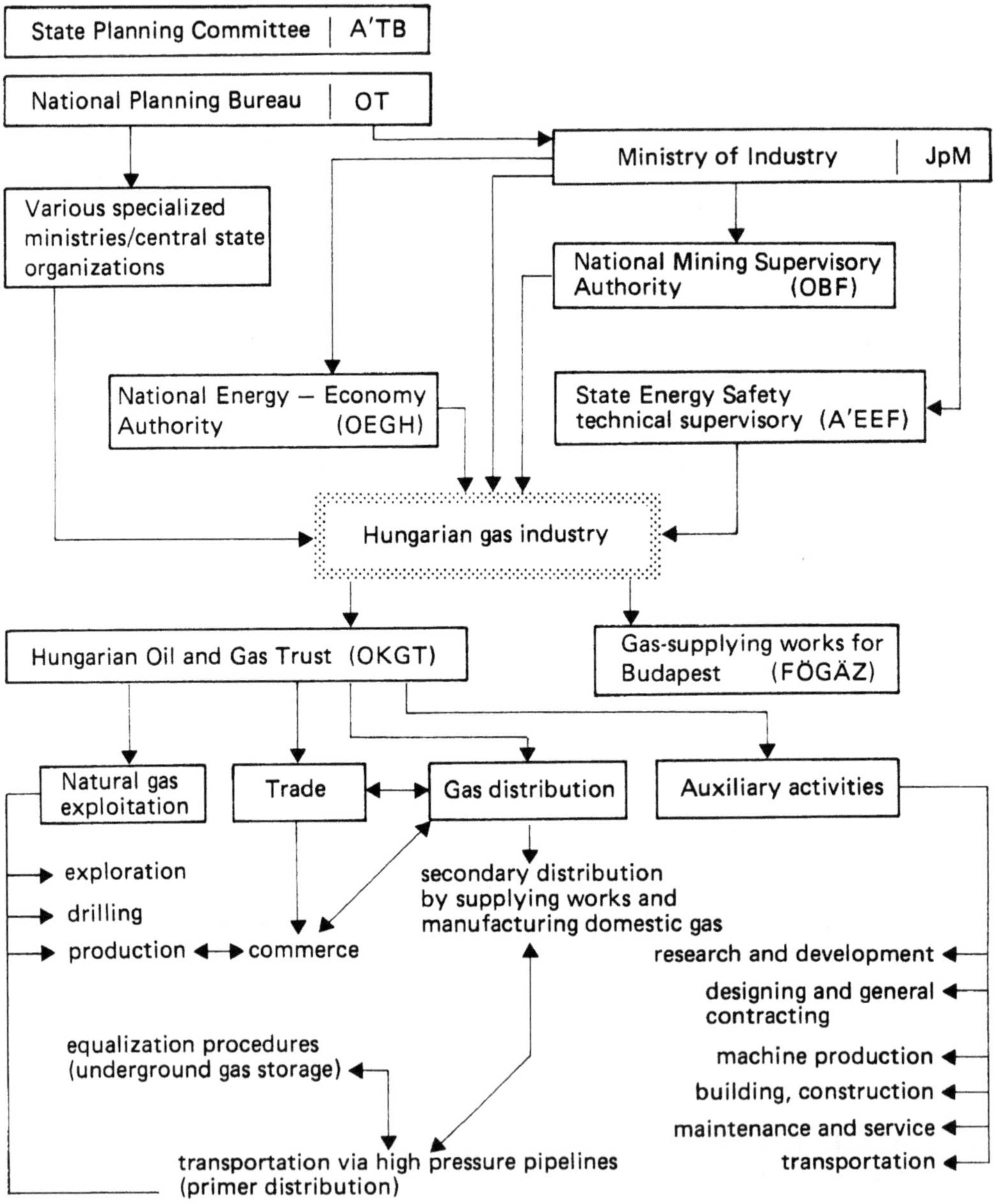

Figure 13.6. Organization of the Hungarian gas industry.

products, using both domestically produced and imported gas. *Figure 13.6* surveys the whole complex system.

Considering the technological (*Figure 13.5*) and management (*Figure 13.6*) conditions, it is a most complex processing system. Its optimum operation affects, directly and indirectly, almost every sphere of economic life. This involves great responsibility, and requires effective coordination and computer communications.

Table 13.2 depicts the significance of natural gas for the national economy and provides information on the development trend. It shows the significance of hydrocarbons in the country's energy supply, with the role of oil decreasing and that of natural gas increasing. Natural gas is also used in significant quantities to produce electric energy.

Table 13.2. Hungarian sources of energy (%).

	1950	*1955*	*1960*	*1965*	*1970*	*1975*	*1980*	*1985*
Energy sources:								
Coal	66	66	68	62	48	32	25	22
Oil	8	16	19	22	29	38	36	32
Gas	4	3	3	6	14	19	27	30
Other	22	15	10	10	9	11	12	16
Total	100	100	100	100	100	100	100	100
Fuel sources for electricity:								
Coal			94		66		51	47
Oil			6		19		14	11
Gas			–		15		35	33
Nuclear			–		–		–	9
Total			100		100		100	100

Natural gas started to play an important role in 1965, and has been gaining in importance ever since in the Hungarian energy management system. (See the increase in the production of natural gas in *Figure 13.1*.) That rapid development was due to the exploration of significant new gas resources between 1960 and 1965, based on a special government energy program (the "Gas-program"). The successful execution of that program is reflected in *Figure 13.7*, displaying the restructuring of gas utilization in Hungary between 1965 and 1990.

Taking into account *Table 13.2* and *Figure 13.7* and the conclusions drawn from these data, it is no exaggeration to say that Hungarian energy management has entered "The Methane Age". We consider it to be of special significance that there has been a very rapid increase in gas consumption by communal consumers and households, as shown by *Table 13.3*. The increasing gas use of these consumers greatly influences the country's social and cultural standards.

With domestic production capacities lagging behind the growing demands, the Hungarian natural gas industry has been compelled to enter the international market. As a result, there has been a gradual increase in imports of natural gas – see *Figure 13.8*. In addition to importing gas, the

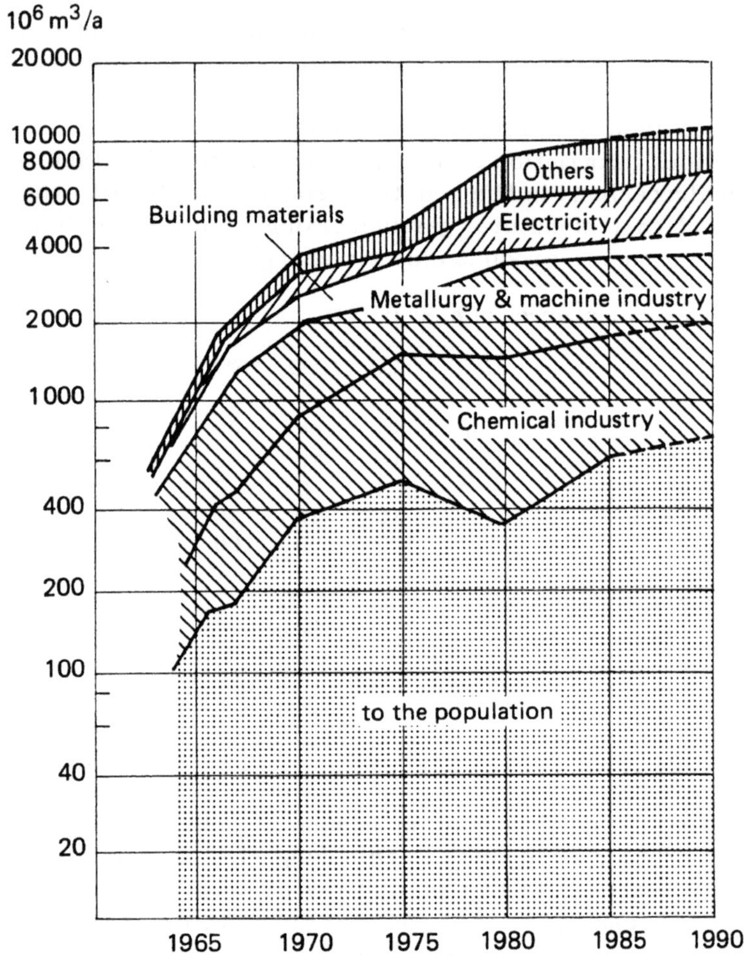

Figure 13.7. Natural gas consumers, 1965–1990.

Hungarian gas industry undertook the task of operating transit gas lines, and expects to handle an increasing volume. The Hungarian hydrocarbon mining industry offers its services on the international market in other fields as well, such as underground gas storage on a contract basis, transit storage of various gas products, different related assignments in foreign countries (such as drilling; seismic prospecting; construction of long-distance lines, telecommunication and telemechanic systems, and reservoirs; engineering and project design; production of various materials, implements, and their export; etc.).

The significant proportion of natural gas in the energy system calls for an efficient, economic gas management activity. Moreover, special developments in the world market price of hydrocarbons made us reevaluate our previous energy policy concepts, and encouraged us to create a new energy

Table 13.3. The number of flats and settlements supplied with gas: 1960–1990.[a]

Year	Settlements connected to the gas system at year-end	No. of flats (× 1000) supplied at year-end			
		Natural gas	Domestic gas	Propane–butane	Summarized
1960	38	9	293	129	431
1965	40	22	376	284	682
1970	78	131	393	1,291	1,815
1971	86	167	392	1,319	1,878
1972	93	229	387	1,383	1,975
1973	99	248	380	1,439	2,067
1974	100	296	370	1,528	2,194
1975	106	350	357	1,623	2,330
1976	110	421	327	1,741	2,489
1977	118	501	298	1,872	2,671
1978	128	562	275	2,003	2,840
1979	140	600	270	2,100	2,970
1980	153	620	267	2,200	3,244
1981	168	707	253	2,259	3,219
1982	186	761	240	2,333	3,334
1983	232	841	207	2,387	3,435
1984	290	935	168	2,440	3,543
1985	324	1,055	127	2,500	3,682
1986–1990	424	1,430	–	2,750	4,180

[a]The data refer to the whole country. The 1986–1990 data are estimated.

policy. Accordingly, the main features of the new natural gas management policy are as follows:

(1) Energy saving and, compared with earlier periods, temporarily more limited investment–development possibilities, changing the production structure in favor of processes that require less energy and, consequently, slower growth in energy consumption.
(2) Changes in the energy structure, mainly replacing crude oil with coal and nuclear energy.
(3) Optimum utilization of the available natural gas, in terms of its efficiency, environmental advantages, and technological superiority.
(4) Optimum peak load management.

We should also mention some borderline issues of effective gas management, such as:

(1) Execution of energy-saving concepts related to the heating and air conditioning of the premises (such as the reduction of heat losses, improving insulating efficiency, combined modes of operation with higher efficiency by utilizing auxiliary energy resources, improving the efficiency of combustion and heat transfer, automatic flue gas

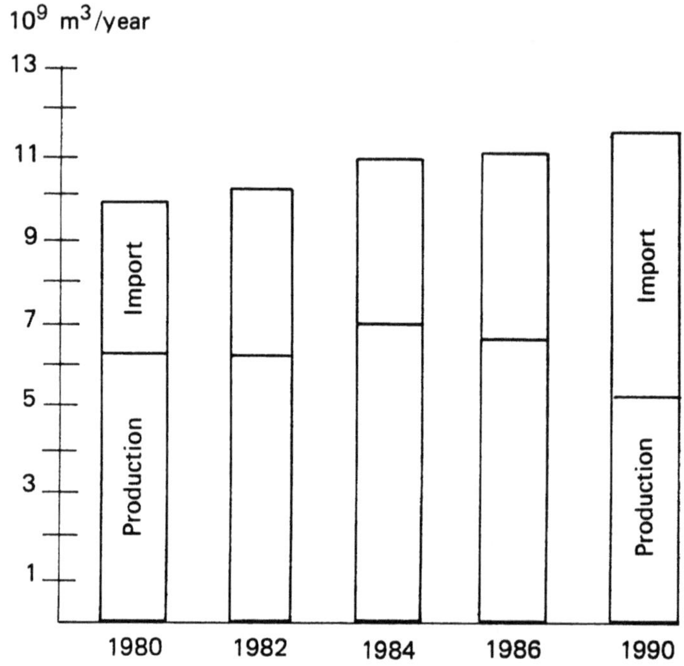

Figure 13.8. Natural gas production and imports, 1980–1990.

transport, utilization of waste heat, the application of automatic facilities, and making discontinuous heating possible).

(2) Development of the instruments and implements utilizing primary energy (e.g., increasing heat recovery, forced ventilation, microwave energy, waste heat, etc.).

(3) Increase in the technical and commercial development of means and materials utilizing secondary energy (e.g., baking in foils, oven-ready dishes, precooked dishes, Teflon-covered pans, etc.).

(4) Emphasis on the development potential of new appliances and helping to get them to consumers (e.g., dip-in and infrared lamps; tubed water heaters for household use; "cold operation" heaters working on the principle of catalytic decomposition; combined refrigerator–water heater systems; complex systems for heating, humidifying, hot water supply, etc.).

(5) Development and popularization of implements operating on the principle of the heat pump.

All these concepts are mainly aimed at saving energy.

13.5. Special Problems of the Natural Gas Industry in Hungary

Figure 13.9 shows that Hungarian-produced natural gas contains a high percentage of natural CO_2 and also has a high inert gas content. Moreover, in the category of gases with high combustibility, the percentage of lean mixtures – that is, gases that require limited processing – is decreasing at the expense of gases that need more elaborate treatment (see *Figure 13.10*). *Table 13.4* refers to the interrelation between production and quality. The technical problems related to the size and qualities of the reserves are discussed in turn below.

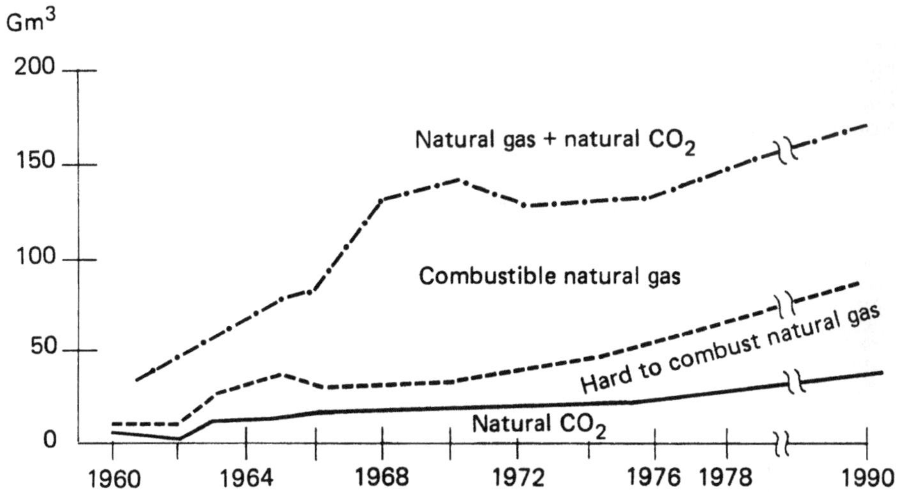

Figure 13.9. Composition of domestic gas reserves.

13.5.1. Gas quality

Hungary's comparatively small system (the country is not large) is fed by almost 50 gas qualities. The composition of the gases varies, and they have to be processed differently for long-distance transportation. As a result, and with special regard to small-sized fields and to the proportion of gases with large inert content, it is very difficult to guarantee permanent gas quality in time and in space.

The problem of maintaining the calorific value within fixed limits has been solved, but it is almost impossible to retain a constant Wobbe number, due to the changes of components. This creates problems for consumers with gas appliances and for the natural gas component separators. In the case of small fields, it is a particular problem to separate heavy

Table 13.4. Classification of Hungarian natural gas (%).[a]

Category	Percentage
I. Calorific value	
1. Combustible gases (>31 MJ/m^3)	54.0
2. Gases of low combustibility (5–31 MJ/m^3)	23.7
3. Noncombustible gases (<5 MJ/m^3)	2.6
4. CO_2 (min. 90 vol. % CO_2)	19.7
Total	100.0
II. Reserves	
1. Small reserves (<200 Mm^3)	1.5
2. Medium-sized reserves (200–400 Mm^3)	2.0
3. Large reserves (>400 Mm^3)	74.2
4. Noncombustible + CO_2 reserves, total	22.3
Total	100.0
III. Production phase	
1. Field under exploitation	54.2
2. Field to be exploited	23.5
3. Noncombustible + CO_2 reserves	22.3
Total	100.0

[a]Data to January 1981. Part of the CO_2 reserves are being exploited.

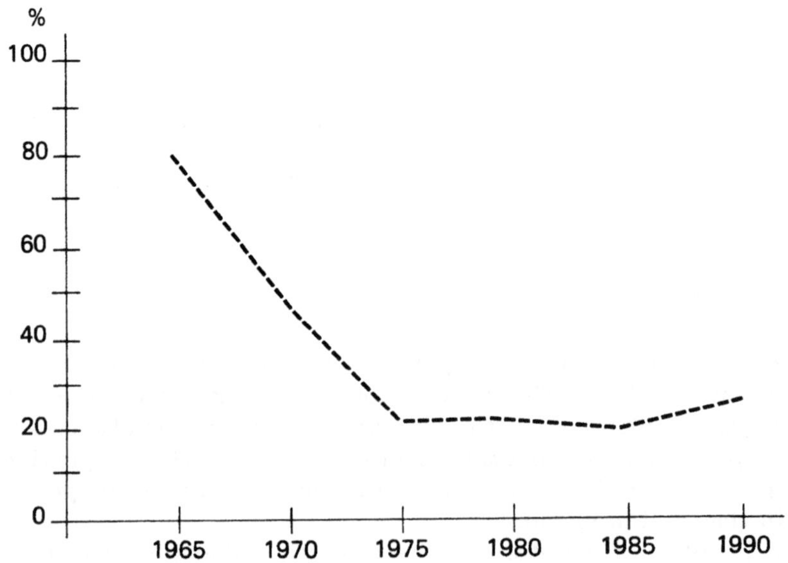

Figure 13.10. Lean gas as a percentage of all natural gas production, 1965–1990.

components, which adversely effect the average Wobbe number of the gas to be transported through long-distance pipes, as well as the combustion properties.

13.5.2. Processing natural gas gained from new wells

The production of natural gas from known wells and from predicted ones must start quickly. Considering the small size of the country, and taking into account already existing gas processing centers, for quality reasons the natural gases exploited from the new wells are taken mainly to these centers for processing. This is all the more reasonable, because it is more economical to construct a storage unit and long-distance pipeline, and operate them, than to build a complete new preparatory or processing plant. This solution calls for solving the transport problems with mixed phases – as is the case with transporting raw natural gas – with special regard to paraffin.

13.5.3. Exploiting fields with small reserves

When exploiting fields with small reserves, we have to give preference to local utilization possibilities, which require the development of special gas preparation technologies with low capacity, and remote supervision capability. Should there be no possibility for local utilization, the raw gas has to be transported to an existing processing center in the vicinity.

13.5.4. Utilizing reserves with high inert gas content

Here we have to provide for conditions to utilize noncombustible gases or gases with low combustibility. Their quantities are significant within the existing reserves – see the data on the relevant figures. The utilization possibilities are as follows:

Group I: Automatic mixing cycles must be established to serve the inlet of new gases, which can be introduced into the existing system.

Group II: Enriching technologies must be set up with existing reserves. However, when conducting enriching operations, it must not be forgotten that, in addition to CO_2, N_2 is also present, which is difficult to eliminate.

Group III: Utilization by targeted user – mainly, major industrial enterprises – can be directed by producers of electric energy.

The solutions to these problems include development of environmental protection facilities, aiming at the utilization of the CO_2 gas that is separated from natural gas, or of flue gas.

13.5.5. Utilizing natural CO_2 resources

This is a special Hungarian problem, the solution to which is an important economic task. As a result of research work that is becoming more expensive every year, the country has a wealth of disclosed CO_2 reserves. In practice this wealth is being further supplemented by the CO_2 separated from natural gas in enriching plants. The large quantity of CO_2 gas available can only be utilized as part of a major, complex technology. The utilization possibilities are:

(1) CO_2 recompression for secondary and tertiary oil exploitation.
(2) Sintered magnesite production, using the technology developed in Hungary and applied to industrial production.
(3) Utilization by the food industry (syphon containers, dry ice, and various refrigerating technologies).
(4) Protection of human life and properties (fire protection).
(5) Agricultural utilization (leaf manure, greenhouse utilization, grain storage, etc.).
(6) Various industrial usages (protective gas for welding, high-pressure CO_2 technology in industries with a possibility of explosive dust, etc.).
(7) Utilization in hydrotherapy.

To improve the efficiency of CO_2 utilization, the problems of transportation and preparation must still be solved. Thus, for instance it is a special technical problem for dry gas, which contains 85 to 88 vol. percent of CO_2 gas in supercritical condition, and so on.

13.5.6. Effective operation of gas and precipitate beds

To achieve the expected results, a practical model with adequate precision has to be created with special phase, balance, and thermodynamic characteristics. In this field we can report on an effective, useful, computerized mathematical model that we have elaborated.

It is a special problem to lift the fluid from the well, due to special surface active agents used during the process. A technical task to be solved is protection against corrosion, as gas precipitate beds almost always include CO_2, H_2S, and steam at high temperature. Paraffin also creates problems during processing.

Special processes have to be applied to exploit as fully as possible the resources available, to obtain the required pressure, and to bring the condensed residues to the surface that remain unexploited during the normal process. Under Hungarian conditions, the gas fractions with high inert content can be most economically used for this purpose.

13.5.7. Gas processing

The proportion of gases that need only drying is decreasing year by year (see *Figure 13.11*). Consequently, we expect a significant increase in the production of liquid gas, and a wider range of gas products. Thus, for example, we can produce, from natural gas, ethane, propane, iso- and normal butane, pentane, hexane, and heptane fractions, etc. We plan to produce these compounds and are working on the establishment of their industrial base. Consequently, our development activities are concentrated on subjects, such as:

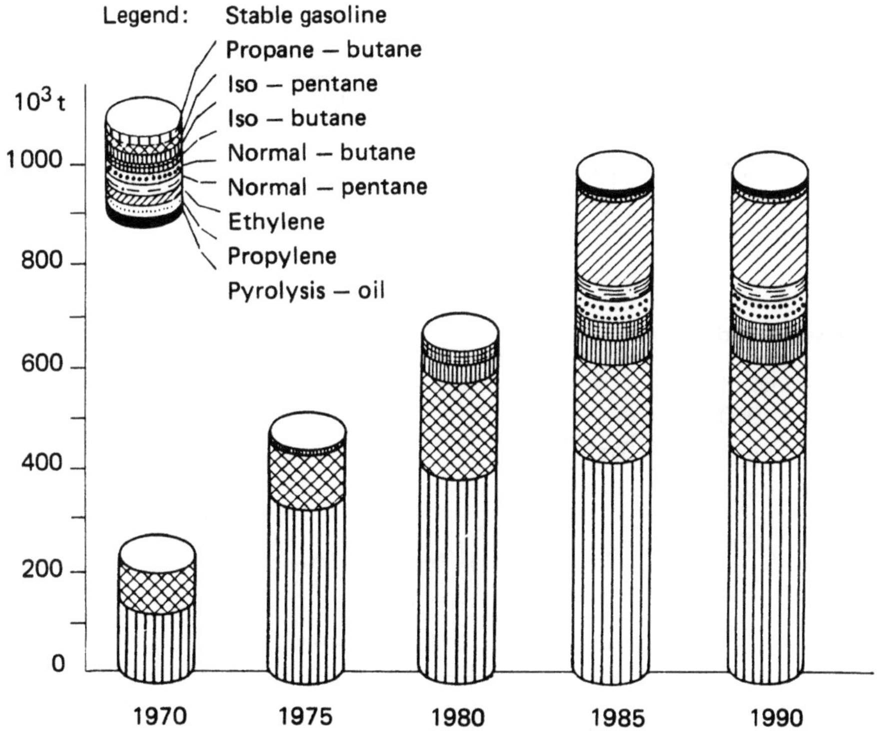

Figure 13.11. **Changing mix of liquefied gas products, 1970–1990.**

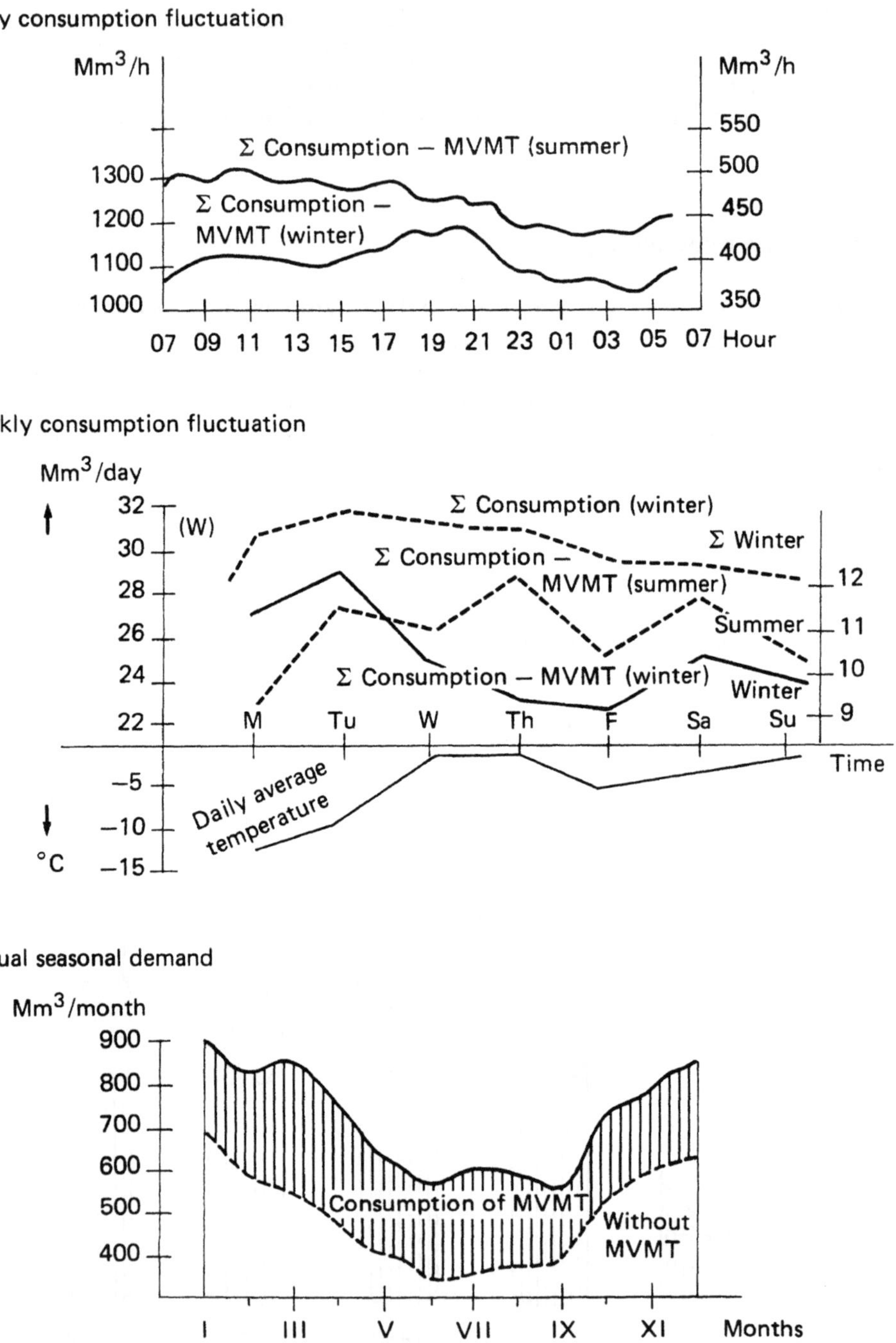

Figure 13.12. **Characteristic consumption fluctuation in Hungary, 1979: (*a*) daily; (*b*) weekly; and (*c*) annually. MVMT = Hungarian Electrical Trust**

(1) Development of highly efficient turbine separators of small volume and weight.
(2) Special spraying technologies to increase the output of the given product.
(3) Use of installations fully incorporating pressure energy.
(4) Extending the product range while using existing installations.

13.5.8. Peak management

Starting out from the changing consumer structure, as presented by *Figure 13.7*, and adapted to the increasing proportion of imported gas, it is an important task for us to establish an optimum peak management system for the natural gas industry. *Figure 13.12* presents the daily, weekly, and yearly fluctuations in the domestic consumption of gas. Fluctuation is on the increase, as an increasing proportion of consumption is related to the weather. This tendency is supported by the data of *Figure 13.13*, and *Figure 13.14* displays the fluctuations in the total of gas consumption.

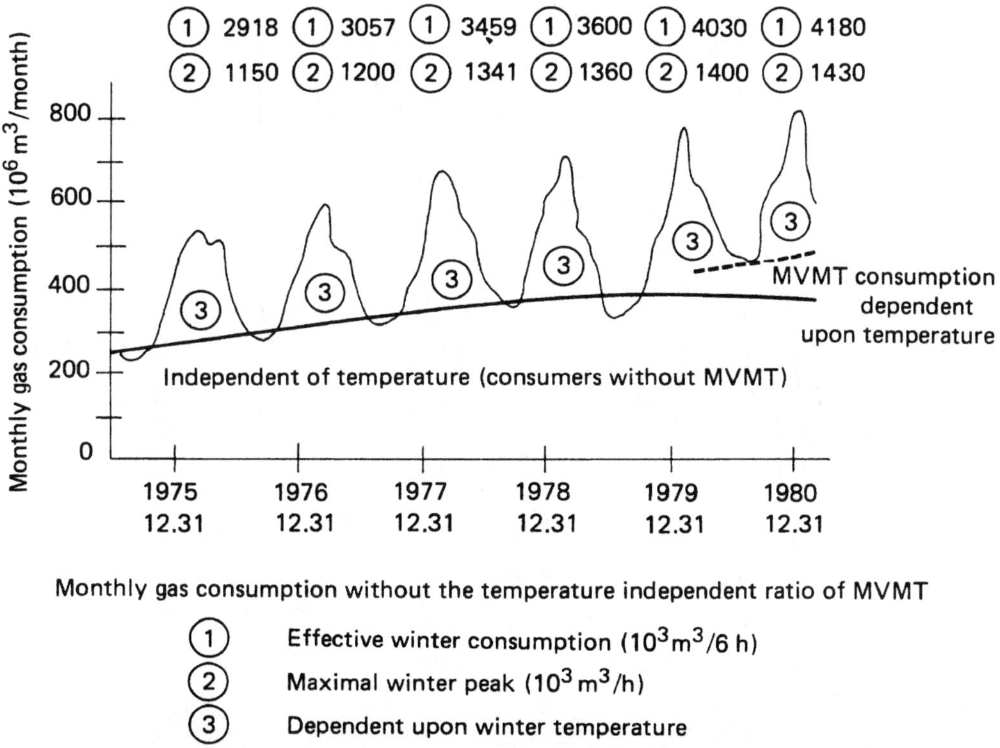

Figure 13.13. **Hungarian gas consumption between October and April, dependent and independent of temperature, 1975–1980.**

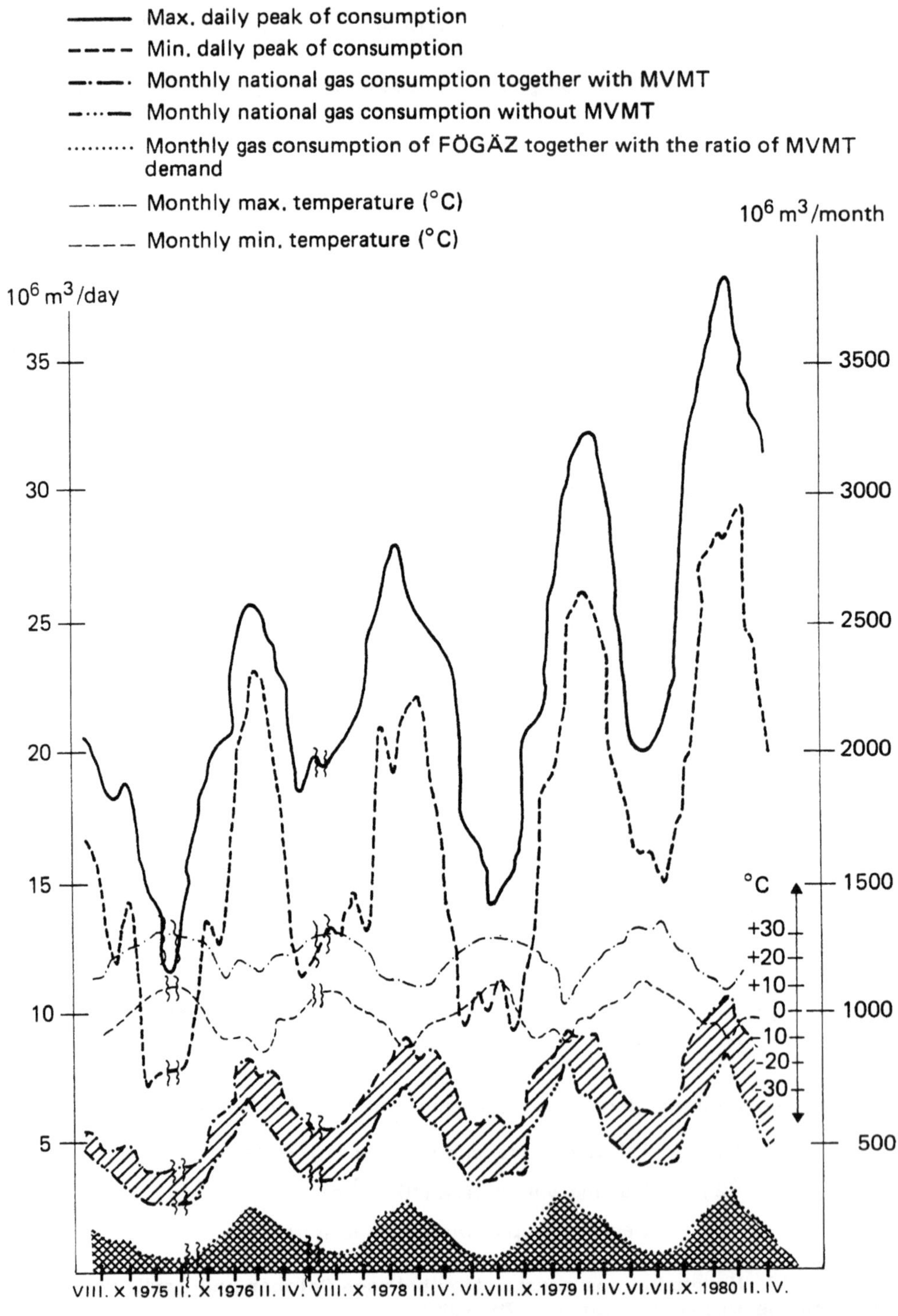

Figure 13.14. **Daily and monthly trends of natural gas consumption, 1975–1980.**

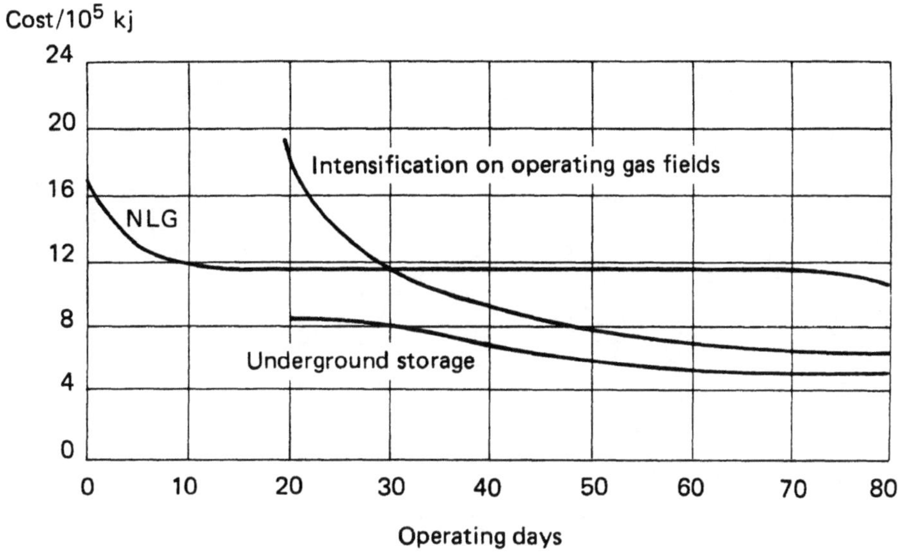

Figure 13.15. **Estimated costs for peak demand equalization.**

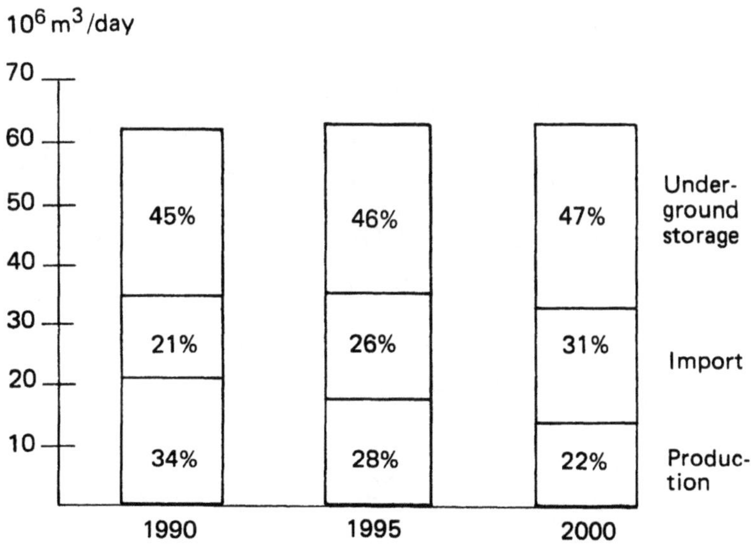

Figure 13.16. **Peak consumption on winter days, 1990–2000.**

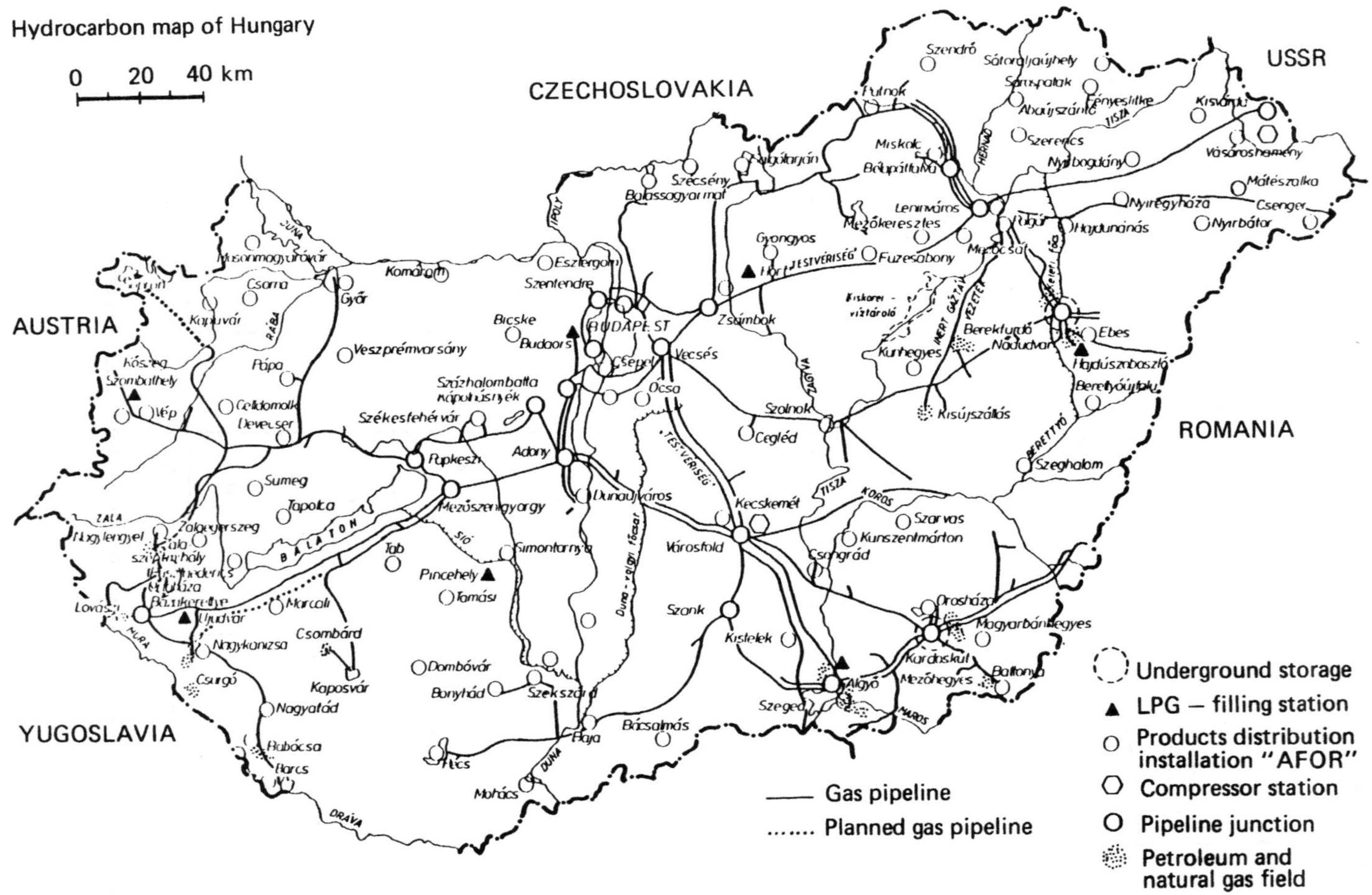

Figure 13.17. **Present and projected gas facilities in Hungary.**

Uneven consumption and gas import requirements demand the construction of underground storage facilities. *Figure 13.15* shows why, economically, we prefer to use underground storage facilities for balancing peak consumption. *Figure 13.16* shows the peak forecast related to the storage capacity development program.

After the entire project is completed, we shall have the capability of storing 8 Gm^3 gas as reserve, an input capacity of 1,720,000m^3/h, an output capacity of 2,920,000m^3/h, and a seasonal storage capacity of 7 Gm^3 gas. This storage capacity will also enable us to store gas on a contract basis.

13.5.9. Transporting via pipelines

Figure 13.17 displays present conditions and plans made for domestic utilization. The operation of the pipeline system calls for the establishment of automated turbocompressor plants, the setting up of an effective and computerized telemetry system, while providing for adequate communication and data transmission systems. Naturally, we had to solve the problems of corrosion protection, adding an odorant to the gas, and continuous monitoring of the quantity and quality of the gas. We have achieved significant results in the fields of defect elimination, pipeline cleaning, etc., and in general maintenance.

13.5.10. Related problems

Of the problems already solved or those we regard as urgent, here are some worth mentioning:

(1) Domestically produced gases tend to have a higher H_2S content, which must be separated before the gas can be utilized.
(2) Because of the high temperature gradient typical of Hungary, it is particularly important to use a well structure technology and techniques that are safe under such conditions.
(3) Gas operating technologies will have greater energy requirements for cooling and pressure increase. This is why special attention must be paid to minimize energy consumption and ensure better use of natural storage energies.
(4) There should be a more extensive use of computer and microprocessor techniques and technologies.

13.6. Summary

In Hungary natural gas has been playing an important role in energy management, and its significance is not expected to decrease in the future. This creates great responsibilities for those working in the natural gas industry who must solve a wide range of problems. Hungarian experts have accumulated significant experience during the past 50 years, as the country has been compelled to make the best possible use of the limited reserves available. We are confident that our experts are fully qualified to respond well to the increasingly complex technological and technical demands of development.

CHAPTER 14

Summary and Conclusions: Likelihood of a Coming Methane Age

H.-H. Rogner, S. Messner, M. Strubegger, and E. Schmidt †

Based on the preceding chapters and discussions at the workshop that generated this volume (see the Foreword), this chapter summarizes the key questions addressed and the conclusions and recommendations that emerged.

14.1. Methane as the Major Global Fuel

With one exception (see section 14.5 below), the authors agreed that methane will probably continue its historically observed market penetration, which began at the turn of the twentieth century. Two forces determine the extent and pace of methane's future market penetration: its availability (supply) and environmental considerations in relation to alternative fuel sources.

14.1.1. Availability

Far more competitively priced methane is being found today than was previously thought to exist. New worldwide reserves, especially outside the USA, are expanding rapidly (see *Figure 14.1*). Furthermore, methane is now found to be more broadly distributed around the earth, both on- and offshore, than oil. This makes it a much more attractive fuel from the standpoint of geopolitics, national security, and balance of payments.

If Dr. Marchetti's logistic substitution model and forecasts prove to be true (see Chapter 3), the amount of methane to be found must be enormous. The analyses given at the meeting showed that methane demand would reach

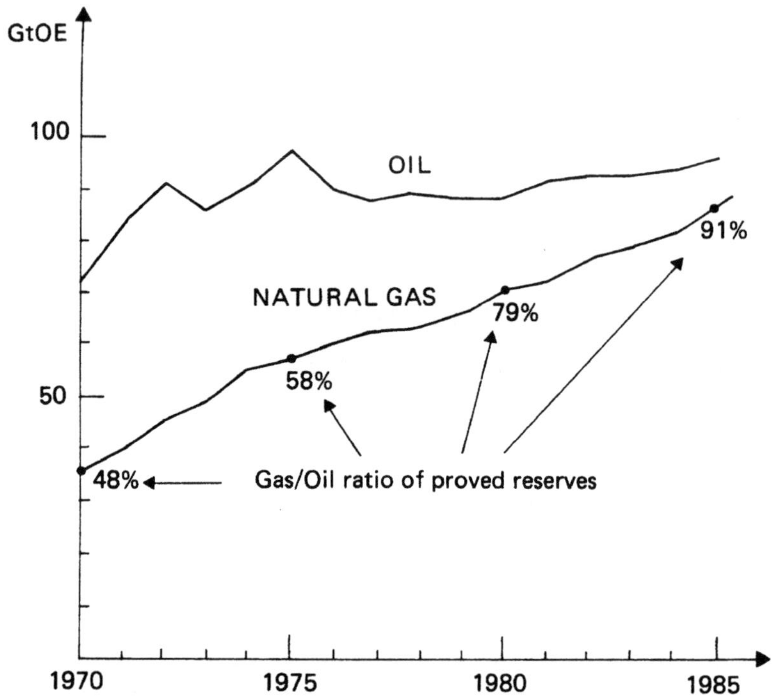

Figure 14.1. Evolution of proven reserves of oil and natural gas in the world (Source: Cedigaz, 1985).

its peak around the decade 2060–2070 at a level of about 30 trillion cubic meters (tcm). This is roughly an order of magnitude larger than present world oil consumption in energy terms, and roughly 12 times the present world consumption of 1.7 tcm of methane. The asymptotic level for integrated methane consumption is about 2,500 tcm, of which less than 2% has already been extracted.

The new super-scale dimension of the methane resource issue is obvious. If these quantities are to be realized, completely new sources of methane must be found and developed.

In that connection, an abiogenic origin for these required new sources is postulated by Professor Gold (Chapter 5) – an hypothesis startling, provocative, and yet to be proven, even though suggested by several Russian, German, and French scientists in the nineteenth century. In essence, Gold integrates the present understanding of the earth's formation process with the abundance of hydrocarbons in the solar system and concludes that most hydrocarbons are the result of the outgasing of abiogenic methane trapped early and deep within the earth.

The theory of the nonbiological or abiogenic formation of most hydrocarbons is gaining credibility and increasing worldwide acceptance. Current drilling in the Siljan Crater in Sweden is one attempt to confirm the existence of such resources, an effort financed by the Swedish State Power Board and the US Gas Research Institute. However, drastic advances in exploration and drilling technology may be required to produce abiogenic methane at competitive prices.

Dr. Hanneman (Chapter 7) characterizes the methane resource situation in more detail, listing such future methane source options as tight sands, Devonian shales, coal seam methane, pure pressure aquifers, gas hydrates, subducted deep sources, and other abiogenic sources. He estimates the unconventional resources of the USA alone at over 8,500 trillion cubic feet (230 tcm). This compares with current official reserve figures of less than 200 trillion cubic feet (5.4 tcm), and current US annual consumption of about 17 trillion cubic feet (or 460 billion cubic meters). One should note, however, that these vast unconventional resources are largely unrecoverable with today's technology.

14.1.2. Environment

Methane is universally acknowledged to be the most benign available fuel by a wide margin. This is especially true if economic comparisons are made on an expanded basis that includes societal and environmental damage effects as well as the usual standard utility accounting components of fuel costs, capital costs, and operating and maintenance costs. Specifically, the external damage caused by burning a ton of coal or oil are estimated to be over $100. The corresponding costs for methane are estimated to be about $35 (see Working Consultative Group..., forthcoming; and Chapter 11 in this volume).

As the anti-nuclear movement grows and continues to block the construction of new facilities or force the shutdown of already existing plants, methane offers itself as a readily available and acceptable replacement fuel. (See also section 14.6 below).

14.2. Innovation and Improvement in Methane Technologies

Compared with historically observed technical progress in related fields, informal estimates suggest that overall technological improvement in the full set of methane-related technologies in the next 50 years could be as great as the overall performance improvement in commercial aviation in the past 50 years. Clearly, these estimates are based on lifetime experience and engineering judgment rather than on arithmetic.

Several methane-related technologies were discussed in intriguing detail in Chapters 6 (Varnado), 9 (Lee), and 10 (Schur and Metz). Those likely to have the greatest potential impact on overall productivity growth include:

(1) Satellite search.
(2) Magnetic anamolies: super-sensitive aeromagnetic mapping.
(3) Gravimetric and density anamolies.
(4) Carbon isotope ratios.
(5) Helium and helium isotope ratios.
(6) Radon excesses.
(7) Selective metallic element anomalies.
(8) Vegetative anomalies.
(9) Electrochemical anomalies.
(10) Fully computerized geochemical/geophysical prospecting.
(11) Searching for lithogenic as well as abiogenic methane.
(12) Searching for methane completely unassociated with oil or sedimentary deposits.
(13) Vastly improved 3-D seismic.
(14) Ultra-deep drilling techniques.
(15) Polycrystallic diamond bits combined with smart downhole electronics.
(16) Wide bandwidth MWD (measurement while drilling).
(17) Vastly improved downhole motors.
(18) Highly steerable drilling packages guided by hardened electronics.
(19) Vastly improved drill pipe and casing materials.
(20) Drilling hundreds of producing holes from a single platform (both off- and on-shore).
(21) Scientific fracturing.
(22) Bigger diameter pipes.
(23) New, highly automated pipe laying techniques.
(24) Building the pipes on site with metal ion/plasma spray techniques (Sandvick process).
(25) Substantial improvements in the efficiency of pipeline pumping turbo machinery.
(26) Improving efficiency of gas turbine/combined cycles from current 47% to achievable 60% and potential 65%

14.3. The Pricetag for Security: Transmission and Distribution Costs

Does it pay to spend several times more on ultra-deep drilling for methane in Europe's heartland to insure the safe transmission of fuel to its market? Those who have invested in the ultra-high-cost producing areas of Troll and Sleipner apparently believe the answer is yes – cheaper over the long term than producing on the periphery of Europe and paying high costs in transportation and anxiety about dependable deliveries (see Chapter 1).

Some experts have raised the point that, to date, no one has carefully disaggregated the component costs in the cradle-to-grave process of finding, developing, and delivering methane to its many different end-users in Europe. In response to this suggestion, IIASA has produced some first-order effect calculations for heartland Europe. Preliminary results clearly show that even though the North Sea exploration and production costs were extremely high (many billions of dollars for a single platform), the dominant cost component to the European end-user was the pipeline (Strubegger and Messner, 1986).

By the same token, activities such as the FRG's Scientific Deep Drilling Program (DM 450 million), the USSR's well established Scientific Deep Drilling Program (Kola Peninsula, etc.), and the quite new complex commercial production drilling program in the Paris Basin (with the basin perhaps positioned directly underneath the city) are all on the right track and worthy of close and careful attention.

If one to two orders of magnitude of technological performance improvement could be achieved, several meeting participants concluded that methane will certainly be the bridge fuel between the first and second nuclear age (see section 14.6). During that period, methane would continue steadily to replace oil and coal (and, in some cases, nuclear fuel) for many, but certainly not all, users. For a recent, well-documented institutional assessment of the possible performance improvement factors that should be considered here, and the RDT&E (research, development, testing, and engineering) funding requirements, see Linden (1985).

14.4. Methane versus Electricity – Fierce End-Use Competition?

Proceeding on the assumption that abundant methane is, and will continue to be, available at attractive prices, Dr. Lönnroth raised this very interesting issue as to the end-use competition between networked electricity generated from methane versus networked methane as an end-use fuel (Chapter 12).

A first-order assessment for heartland Europe suggests that the size of the generating stations would shrink by one to three orders of magnitude and be much more broadly dispersed. Another possible net result would be that more power would be distributed by a finely structured electric grid (fueled by a moderately coarsely structured methane grid) rather than by a super-finely structured methane grid (Lönnroth and Schmidt, 1986).

In the USA, the results of the EGEAS (Electric Generation Expansion Analysis System) Study undertaken at MIT were unexpected (EPRI, 1982). They indicate that methane-fired combined cycles would clearly be the most cost-efficient method of new electric generation installations in seven out of nine utility districts making up the entire US generation grid. The EGEAS study concluded with the following statement, referring to natural gas-fired combined cycles: "Their economic competitiveness cannot be denied. Their environmental advantages, safety aspects, ease of siting, and modularity make them an ideal system in an era of uncertainty."

It is most important to note here that we have no generally accepted method for measuring "attractiveness" of an end-user system, or a primary energy source. The old measures, i.e., cost per kilowatt hour of electricity generation, wholly based on input costs, have been highly misleading. In some cases they have led to catastrophic financial and societal results, to the point of imperiling the future of several new technologies and the commercial atomic power industry. Thus, several workshop participants viewed the development of an entirely new set of attractiveness measures as perhaps the primary task for system-oriented researchers and practitioners over the next decades.

Apart from the familiar simplistic terms relating to straight capital, fuel, and operating and maintenance costs, more complete and realistic measures should account for:

(1) General societal acceptability of the entire system.
(2) Total external damages caused by installing and operating the system on a cradle-to-grave basis.
(3) General robustness of the system to all forms of possible interruptions.
(4) Minimum lumpiness of capital and physical plant investment.
(5) Minimum optical comprehensibility per systems unit (systems that can hardly be seen are liable to be evaluated as much more attractive than those with high optical comprehensibility).

No real substantive work has been done yet on developing a true attractiveness measure for energy systems. We suspect that any such measure will be organized in the form of rather complex functions with many heavily weighted value terms representing comprehensive societal costs.

14.5. Methane: A Provincial, International, or Intercontinental Fuel?

Dr. Dreyfus (Chapter 8) heavily stressed the point that the future role of methane is far from clear. In the USA, methane tends to be mostly a national or even regional fuel (in Dreyfus's terminology, a provincial fuel). Currently, in Europe methane is an international fuel. In Japan, it is a truly intercontinental fuel based almost solely on LNG transport. The fundamental issue seems to be whether or not transportation costs (particularly LNG costs) can ever be sufficiently lowered to make methane a truly intercontinentally traded commodity.

The fact is that LNG in general and Japanese LNG in particular are still markedly uneconomic. Apparently, cost considerations are not the only criteria for the Japanese, because their imports of LNG have increased from 3.1 billion cubic meters (2.8 million tons of oil equivalent) in 1973 to 37.6 billions of cubic meters (34 million tons of oil equivalent) in 1985. This is over a twelvefold increase in 12 years. Still, to put matters into perspective, world LNG transport, most of which is to Japan, amounts to just 3% of total world methane use (OECD/IAEA, 1986).

14.6. Methane versus Nuclear Power

First-generation commercial atomic power plants and all related nuclear technology development appear to be stalled in most of the West for at least several decades. For example:

(1) Austria has built a commercial atomic power plant that the electorate voted not to start. The physical removal (full site clearing) of this mint-condition plant has been decreed.
(2) Sweden's anti-nuclear majority has compelled the authorities to schedule the shutdown of its entire fleet of commercial reactors. Anti-nuclear public opinion certainly appears to be increasing in most Western countries with the possible exception of France.
(3) In the USA, the world's major nuclear power operator, cancellations of multibillion dollar power stations, frequently over 75% complete (and sometimes as high as 95% complete), are no longer uncommon. The most recent cancellation – the $4 billion Consumers Power Midland Station (95% complete) – will be converted to a methane-fired, combined cycle unit. The methane will most probably come from nearby high-flow-rate wells that have been recently completed.

Investigations are now being made of the effects of shutting down all or major portions of the Western world's commercial reactor fleet. An analysis made by the Rand Corporation (see Stucker, 1985), based on the shutdown of several site-specific reactors, concluded that the net effects on local consumers and surrounding industry would be almost unnoticeable. Six independent West German studies came to similar conclusions. IIASA has made one such first-order-effects analysis for a nuclear phaseout in Central Europe (Messner and Strubegger, 1986). Two of the significant conclusions are especially relevant here:

(1) In heartland Europe, if a Swedish-style commercial atomic power moratorium were declared today, the macro-effects on the consumer in particular and industry in general would be entirely manageable.
(2) The full effect of a phaseout would occur after the turn of the century. Much of the networked replacement energy would come from new electricity generated by methane-fired combined cycle plants or from additional directly pipelined methane. Most simply stated: methane would be the major replacement fuel because it is plentiful, at an acceptable price, and carries acceptable environmental risks.

Thus, while first-generation nuclear energy fades, methane-related technologies and commercial applications are entering a rapid growth phase. Methane may well become the bridge fuel between the first and second nuclear ages.

The term "second nuclear age" must be defined as broadly as possible, including nuclear energy supplied by the solar system. Perhaps it should be called the "second energy age." Specifically, it is consistent with the broadly sketched interfuel substitution patterns covered in *Energy in a Finite World* (Häfele *et al.*, 1981). A second nuclear age is anticipated because of growing environmental concerns regarding the potential adverse impacts of a continual CO_2 accumulation in the atmosphere. Though substantially lower than in the case of coal, the methane age is inextricably associated with CO_2 emissions, unless fossil fuel combustion processes occur in a totally enclosed system and with pure oxygen (Häfele *et al.*, 1986).

14.7. Recommendations for Further Research

The final session of the Sopron workshop was devoted to conclusions and recommendations for further work at IIASA, including:

(1) A case study of increased methane use as a part of the field of the dynamics and impacts of new technology.
(2) A study of the role of countries endowed with large gas reserves (Algeria, Australia, the Near and Middle East in general) and the impact of increased gas trade on world trade, national security, and geopolitics. This study would require a systems approach because both the technology and methane as an energy source act in an international political and economic environment.
(3) Establishment of a methane resource data base with information from those who consider gas separate from oil.
(4) A study of the technology–economy relationship in methane exploration and use. Gas producers have indicated that they would appreciate an economic assessment of the whole "causal technological chain" from drilling to burners. This would constitute an important input for decision making in the exploration area. The tradeoff between local deep drilling costs and long-distance transportation costs must be clearly established. Trends in major cost elements should be identified over a 20–40 year span and their effects disaggregated. This should show what the potential is for methane price reduction.
(5) Investigation of the need for new policy measures to secure cooperation between a gas network and "companion" electric power network.

References

Cedigaz (1985), *Natural Gas in the World in 1984* (Rueil Malmaison, France).

EPRI (1982), *EGEAS: Electric Generation Expansion Analysis System*, RP 1429 (Electric Power Research Institute, Palo Alto, CA.).

Häfele, W. *et al.* (1981), ***Energy in a Finite World – A Global Systems Analysis.*** Report of the Energy Systems Program Group of the International Institute for Applied Systems Analysis (Ballinger, Cambridge, MA).

Häfele, W., Barnert, H., Messner, S., Strubegger, M., with Anderer, J. (1986), The concept of novel horizontally integrated energy systems: the case of zero emissions. In W.C. Clark and R.E. Munn, (Eds.), *Sustainable Development of the Biosphere* (Cambridge University Press, Cambridge, UK).

Linden, H.R. (1986), *Impact of Advances in Science, Technology and in the Terrestrial Origin of Hydrocarbons on the Role of Natural Gas and Crude Oil in Meeting Future Primary Energy Needs* (Gas Research Institute, Chicago).

Lönnroth, M. and Schmidt, E. (1986), *The Coming Age of Gas and Electricity: Scales, Technologies, and Social Organizations*, internal paper (International Institute for Applied Systems Analysis, Laxenburg, Austria).

Marchetti, C. and Nakicenovic, N. (1979), *The Dynamics of Energy Systems and the Logistic Substitution Model*, RR-79-13 (International Institute for Applied Systems Analysis, Laxenburg, Austria).

Messner, S. and Strubegger, M. (1986), *First Order Effects of a Nuclear Moratorium in Central Europe*, internal paper (International Institute for Applied Systems Analysis, Laxenburg, Austria).

OECD/IAEA (1986), *Natural Gas Prospects* (International Atomic Energy Agency, Paris).

Strubegger, M. and Messner, S. (1986), *The Influence of Technological Changes on the Cost of Gas Supply*, WP-86-38 (International Institute for Applied Systems Analysis, Laxenburg, Austria).

Stucker, J.P. (1985), *The Closure of Nuclear Power Plants*, N-2179-RC (Rand Corporation, Santa Monica, CA).

Working Consultative Group of the President of the USSR Academy of Sciences on Long-Term Energy Forecasting (forthcoming), *The International Natural Gas Market* (International Institute for Applied Systems Analysis, Laxenburg, Austria).

Task Force Meeting on
The Methane Age
May 14–16, 1986 – Sopron, Hungary

List of Participants

Csako, Dr. Denes
Head of Department
Hungarian Oil and Gas Trust (OKGT)
Schonherz Z. U. 16
P.O. Box 270
H-1117 Budapest
Hungary

Dreyfus, Dr. Daniel A.
Vice President, Strategic Analysis
and Energy Forecasting
Gas Research Institute
1331 Pennsylvania Avenue, N.W.,
Suite 730 North
Washington, DC 20004-1703
USA

Gold, Professor Thomas
Cornell University
Center of Radiophysics
and Space Research
Space Sciences Building
Ithaca, NY 14853
USA
or: 36 Madingly Road
Cambridge CB3 OEX
United Kingdom

Golovin, Dr. Anatoli
IIASA

Hanneman, Dr. Rodney E.
Vice President & Chief
Technical Officer
Reynolds Metals Company
6601 West Broad Street
Richmond, VA 23261
USA

Hefner, Mr. Robert A., III
GHK Company
6441 North West Grand Blvd.
Oklahoma City, OK 731166
USA

Kaftanov, Professor Vitali
IIASA

Khorkov, Dr. Alexander
Working Consultative Group
on Long-Term Energy Forecasting
of the President of the
Academy of Sciences of the USSR
Vavilov Str. 44, Bl. 2
Moscow
USSR

Lee, Professor Thomas H.
Director, IIASA

Lönnroth, Dr. Måns
Prime Minister's Office
Statsradsberedningen
S-103 33 Stockholm
Sweden

Macak, Professor Jiri
Chief of Department
Gas and Coke Technology
Institute of Chemical Technology
Suchbatarova 5
CS-16628 Prague 6
Czechoslovakia

Marchetti, Dr. Cesare
IIASA

Messner, Dipl.-Ing. Sabine
IIASA

Metz, Dipl. Ing. Jack R.
New Products and Processes
Sulzer Bros. Ltd.
P.O. Box
CH-8501 Winterthur
Switzerland

Nakicenovic, Dr. Nebojsa
IIASA

Rogner, Dr. Hans-Holger
IIASA

Ruttenstorfer, Dr. Wolfgang
Unternehmensplanung
Referat Energiepolitik
OeMV AG
Otto Wagner-Platz 5
1090 Vienna
Austria

Schmidt, Mr. Ed
IIASA

Schur, Mr. Tom
Manager
Task Force Corporate Marketing
Sulzer Bros. Ltd.
P.O. Box
CH-8401 Winterthur
Switzerland

Subai, Mr. Jozsef
Head of Department
General Contracting and Designing
Office for the Oil Industry
Schonherz Z. U. 16
P.O. Box 270
H-1117 Budapest
Hungary

Varnado, Dr. Samuel G.
General Manager
Titan Technologies
2017 Yale Street SE
Albuquerque, NM 87106
USA

Vasko, Professor Tibor
IIASA

Vostry, Dr. Z.
Institute of Information Theory
and Automation
182 08 Prague 8 –Liben
Pod vodàrenshou vezi 4
Czechoslovakia

Zavorka, Dr. J.
Institute of Information Theory
and Automation
18208 Prague 8 –Liben
Pod vodàrenshou vezi 4
Czechoslovakia

The GeoJournal Library

Famine as Geographical Phenomenon, Bruce Currey and Graeme Hugo (eds.), 1984. ISBN 90–277–1762–1.
Environmental Geochemistry and Health, S. H. U. Bowie, F.R.S. and I. Thornton (eds.), 1985. ISBN 90–277–1879–2.
Population Redistribution and Development in South Asia, Leszek A. Kosinski and K. Maudood Elahi (eds.), 1985. ISBN 90–277–1938–1.
Desert Development, Yehuda Gradus (ed.), 1985. ISBN 90–277–2043–6.
Geographical Dimensions of Energy, Frank J. Calzonetti and Barry D. Solomon (eds.), 1985. ISBN 90–277–2061–4.
Strategies for River Basin Management, Jan Lundqvist, Ulrik Lohm and Malin Falkenmark (eds.), 1985. ISBN 90–277–2111–4.
Migration and Settlement. A Multiregional Comparative Study, Andrei Rogers and Frans J. Willekens (eds.), 1986. ISBN 90–277–2119–X.
Spatial Strategies in Retailing, Risto Laulajainen, 1987. ISBN 90–277–2595–0.
The Methane Age, T. H. Lee, H. R. Linden, D. A. Dreyfus and T. Vasko (eds.), 1988. ISBN 90–277–2745–7.
Artificial Structures and Shorelines, H. J. Walker (ed.), 1988. ISBN 90–277–2746–5.

MIX
Papier aus verantwortungsvollen Quellen
Paper from responsible sources
FSC® C105338

If you have any concerns about our products,
you can contact us on
ProductSafety@springernature.com

In case Publisher is established outside the EU,
the EU authorized representative is:
Springer Nature Customer Service Center GmbH
Europaplatz 3, 69115 Heidelberg, Germany

Printed by Libri Plureos GmbH
in Hamburg, Germany